Energized
Cybersecurity
Culture

Energized Cybersecurity Culture

A Marketing Approach To Build Excitement And Increase Participation In Your Security Awareness Program

Bryan Perkola

Bray Publishing

Book Cover by Bryan Perkola

Cover images: szpeti/Shutterstock.com, LightField Studios/Shutterstock.com, Photon photo/Shutterstock.com, Pingingz/Shutterstock.com, CeltStudio/Shutterstock.com

First paperback edition January 2026

ISBN: 979-8-9941482-0-4 (paperback)
ISBN: 979-8-9941482-1-1 (ebook)

Library of Congress Control Number: 2025926884

Published by:
Bray Publishing
Houston, Texas
www.braypublishing.net

Devolutions Sysadminotaur Image © Devolutions Inc. and Patrick Desilets, used with permission.

All trademarks are the property of the respective owners.

To Jenny, Savanna, and Chase, who always provide me support no matter what endeavor I am pursuing. Your unwavering support means the world to me.

About the Author

Bryan Perkola, CISSP, CISM, CDPSE, C|EH, C||HFI is a passionate cybersecurity professional with over 40 years of IT experience in organizations across manufacturing, retail, and finance. Bryan holds bachelor's degrees from the University of Houston in Marketing and Organizational Behavior and Management, and received his master's degree in Cybersecurity and Information Assurance from Western Governors University, in addition to numerous industry certifications.

Bryan's work experience has focused on small- to mid-sized organizations, where he was intimately involved in multiple disciplines, including marketing and human resources, which provided him with unique perspectives and an understanding for developing creative and effective security awareness programs that promote a strong cybersecurity culture within the organization.

In Bryan's current role as Senior VP of Information Security at First Community Credit Union (Houston), he is responsible for the credit union's security operations, including developing a security awareness and training program to engage employees and promote their role in protecting the credit union's data. In promoting cyber-security, Bryan frequently speaks at security conferences, delivers corporate presentations on cybersecurity throughout the year, conducts educational seminars for the credit union's members, and speaks to local high school students about security issues as part of the credit union's commitment to the community.

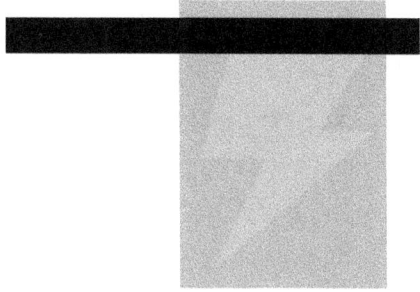

Acknowledgements

This book is a dream realized that I never realized was a dream. I never thought I would write a book in my lifetime, but here is my work, thanks to a little nudge from my daughter, hours behind the keyboard, and the support from numerous people.

Let me start by thanking my family, especially my wife, Jenny, who has put up with me and my endeavors longer than I thought anyone would and has encouraged me in so many ways. Jenny, Savanna, and Chase, you have been the driving force behind so many of my accomplishments, stood by me, and helped me achieve so much. I hope the achievements you have helped me reach inspire you to pursue your dreams and, with the proper support, excel in ways you are not even aware of.

Special thanks to my parents for all the years of love and support. I appreciate all that you have done for me throughout my life.

Nicolle and Adel, I couldn't ask for a better or more committed team to work with. You have worked with me as many of the ideas presented in this book were conceived, implemented, and refined. We have built an outstanding program that continues to get better because of your efforts and contributions. You both have bright futures, and I am glad to be part of your journey.

In this book, I discuss the need for support from the top, and I am fortunate to have that. TJ Tijerina, Sravan Vinjamuri, and previously Rito Garza, thank you for the support and the blank canvas to create and implement a security awareness program based on the principles in this book. Your support is appreciated, and I believe we have validated many of the concepts by the outstanding cyber-security culture we have developed.

Thanks to my friend, Michael-Angelo Zummo, for writing the foreword and providing feedback throughout this process. Zummo was in attendance for the first presentation I gave about creating cybersecurity programs, so he was naturally one of the first people I reached out to when I began this process. I always enjoy our talks, and I appreciate all the guidance you have provided throughout the years.

I want to acknowledge a couple of friends, John Romero and Kyle Hill, who provided valuable feedback and encouragement to continue the process after reading early drafts of this book.

Special nod goes to my friends at KnowBe4. Especially Casey and Hannah, who have given me so many opportunities to present these ideas and strategies for creating an effective security awareness program.

I could not forget to thank Stephen Martin. We share so many common thoughts that it's creepy at times, but when I need a sounding board to work through an issue, there is no one else I prefer to turn to. Your input is invaluable, and I appreciate the opportunity to work with you daily.

Thanks to the entire staff at First Community Credit Union in Houston, Texas. Your active participation in the security awareness program provides proof that an engaged workforce can be an effective component of the security stack. This begins with Andrew, Samantha, and the other members of Staff Development, who introduce new employees to the methodologies of our security program. Special thanks to Amber, Danny, and the entire marketing team for their support and commitment to promoting strong cybersecurity principles to our employees and members.

I have had the opportunity to work with so many wonderful people throughout my career, so here is my blanket "thank you" to all who have shaped my professional growth. I wish I could single out each of you for the contribution you made in helping me develop this program, but that would be another book.

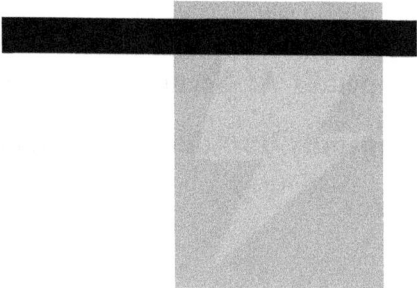

Contents At A Glance

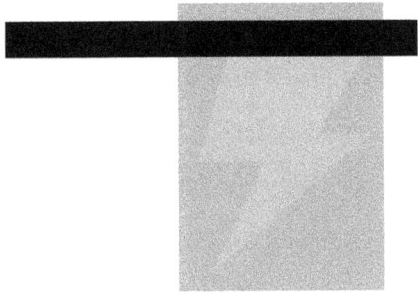

Table of Contents

Foreword

I first met Bryan several years ago when his team was expanding their technical expertise and adding additional security tools. I was responsible for assisting them implement a cyber threat intelligence program and mapping it to their priority intel requirements. I'll never forget my first impression of his team: an incredible sense of pride emanated from each of them.

Over the years, I've been blown away by Bryan's ability to make security fun for an entire organization - well beyond the security department. Weekly phishing challenges are something to look forward to in Bryan's culture - rather than a stress-inducing test. I've watched organizations struggle to turn "awareness" into true cultural change. What

Bryan offers in Energized Cybersecurity Culture is a playbook for doing exactly that. He shows that defending an organization starts not with tools, but with people — and that motivating people requires storytelling, emotion, and purpose.

This book reminds us that cybersecurity isn't a training module; it's a continuous movement that is everyone's responsibility. Bryan's approach, drawn from decades of IT, marketing, and behavioral insight, provides the spark leaders need to make security part of everyday life.

Whether you're a CISO, HR leader, manager, or just someone trying to make security "click" with your team, this book will give you practical, creative ways to build trust, engagement, and lasting change.

Bryan doesn't just teach cybersecurity. He teaches how to make people care about it — and that may be the most powerful defense we have.

Michael-Angelo Zummo
Director of CTI Solution Engineers
Bitsight

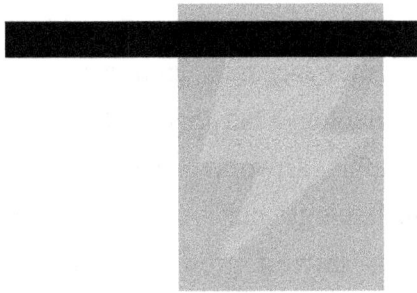

Introduction

The journey to this book…

I believe it is important that I chronicle key events from my background that directly led to the processes and ideas I incorporate into building cybersecurity awareness programs. This look back will provide valuable insight into the development of the skills and methodologies I use to instill a strong cybersecurity culture in my organization.

My career in information technology and journey in cybersecurity began over forty years ago, but it has been anything but a traditional path. In high school, I was heavily involved in the school's drafting department, when computer-aided drafting (CAD) technologies were emerging for personal computers. I spent many hours working on engineering and architectural drawings using this new technology, but then I started experimenting with creating graphic artwork. This was not an ideal use of CAD technology, but it sparked an interest in computer graphics. This interest grew over time and eventually led me to discover and become proficient with graphics applications such as Photoshop and Illustrator. Skills that have been instrumental in creating materials to support my security awareness programs.

At the same time, in high school, because of my interest in computers, I was also asked to create a couple of simple programs for the school that used database applications of the time. This sparked an interest in programming, and over the years, I worked with several languages (Delphi, Visual Basic, and C#). This initial introduction to databases eventually led me to work with enterprise-level SQL databases. I draw on these skills to develop metrics that enable me to drill down into the statistics and data from the security awareness program to evaluate performance.

I then went to college, where I attained two bachelor's degrees. I received one of my undergraduate degrees in Marketing. At the time, I wasn't sure why I chose this path, but in the end, it has been extremely instrumental in helping me understand how to get my employees to participate in the various security programs we present. My second undergraduate degree, Organizational Behavior Management (Human Resources), was a complete accident. I was told I needed a minor, which was twelve hours. Thinking about looking for a job, I thought that if one degree is good, then two must be better. At the time, I thought this would look good on my resume, but it turned out to provide me with invaluable guidance and knowledge in building security awareness programs.

During my time in college, personal computers were starting to enter the workplace, so individuals with computer experience were not in abundance. I had been working with personal computers since high school and continued to do so throughout my college years, working for local companies in various capacities. Due to this strong interest in and knowledge of computers, I naturally gravitated towards working in the information technology field as I entered the workforce.

I began my professional career working with small- to mid-size organizations, where I sometimes had to wear multiple hats and work closely with other departments. This is where my business degrees really helped me: I was able to work very closely with the marketing and human resources departments because I understood their business concerns and operating methods. At times, due to the size of the organization, I performed functions in these departments in addition to my information technology responsibilities, which provided me with invaluable insight I would not have been exposed to otherwise. This insight has been a critical element in my development of cybersecurity culture and awareness programs.

While working with the marketing departments, I had the opportunity to learn from and work with some very talented professionals, who taught me how to effectively develop promotions and artwork to highlight products through coordinated campaigns. My experience with the marketing group and my enjoyment of working with graphic design prompted me to set up a side business in which I created websites and graphic materials for a variety of customers for over twenty years. This side business allowed me to perform contract work for several local printers. This contract work provided me with additional insight into other industries and how they market their products. Working with this diverse group of organizations allowed me to explore and understand how professional marketers were positioning their products for their target markets. This knowledge of target markets and how to position your product became a central focus of my methodology for developing cyber-security programs that rely on visualization, graphics, and coordinated messages to engage employees.

I have been focused on cybersecurity for about twenty years. As I began to progress towards this specialization, I was not even aware I was on that track. While performing my regular IT duties, I started working more in the security space, and one day I realized I was spending most of my time exploring how to attack and defend networks. I really enjoyed that area of IT. In researching the attack and defense of network resources, it quickly became apparent that the human factor often played a significant role in successful attacks, so finding ways to engage my fellow employees in protecting our organization's resources was paramount. The realization of the significant role the human element plays in defending the organization was the catalyst that led me to begin developing cybersecurity awareness programs. As I started developing these programs, I drew on my education and experience working with the marketing and human resources departments to elevate employee participation.

Move to 2016. Over the previous ten years, I had been spending increasingly more time working in the security space, and I began to recognize that the human element plays a significant role in protecting an organization's digital presence. I did not entirely put that together until I started working with platforms that run simulated phishing campaigns and provide additional security training materials. It was at this time that I recognized how the experience I had gained in marketing and human resources could play a key role in promoting the cybersecurity programs to my fellow employees and securing their participation. Over the next several years, I experimented with and refined the marketing elements of my cybersecurity program, which are now at the core of my security awareness programs.

In 2019, I purchased a copy of Perry Carpenter's book, Transformational Security Awareness: What Neuroscientists, Storytellers, and Marketers Can Teach Us About Driving Secure Behaviors. This book solidified many of the principles I had been building into my program, but I lacked the analytical support to fully explain the success of my security program. Perry's book gave credibility to the ideas I had been working on and provided scientific and behavioral support to continue the development of my cybersecurity program. If you have not read Perry's book, Transformational Security Awareness, I highly recommend that you familiarize yourself with the materials. The materials in this book will touch on many of the ideas in Transformation Security Awareness, but will delve deeper into how our program integrated those concepts into a practical application. We will look at some topics from a slightly different perspective and offer unique ways to develop your security awareness program.

Over the next several years, a slew of new issues were introduced affecting the development of security awareness programs. We encountered the pandemic, which led to an increase in the number of employees working from home, presenting several unique challenges in creating a security awareness program that encapsulates the organization's security culture. We experienced new disruptive technologies, such as artificial intelligence (AI), that have altered the workplace and the way employees conduct their daily activities. As these changes unfold, our security awareness programs must continually adapt to provide employees with timely training and methods to ensure the organization's cybersecurity culture remains in focus.

I have provided a summary of my journey in developing cybersecurity awareness programs. What was the impetus for writing this book? I started presenting materials about how I create security awareness programs to influence security culture in my organization at security conferences several years ago. After each presentation, I would have attendees stay to discuss the topics covered with me immediately afterward, or reach out through email or LinkedIn. In either case, the sentiment was always similar: the presentation provided them with direction or vision for getting their program started, or with unique ideas they wanted to implement in their cybersecurity program. I realized from these conversations that many of these individuals lacked a marketing background, so it was not an obvious approach to take when developing a security awareness program. Thus, the inspiration to write this book was born.

Laying the Groundwork

The methodologies outlined in this book have been developed and refined over the last decade and continue to evolve to meet today's challenges. I believe one of the best ways to stay ahead of our common adversaries is to share ideas from others that have proven successful and adapt those ideas into your security program. The exchange of ideas is the premise of this book, and the basis for the presentations I give at security conferences. I hope that through the presentation of the material in this book, you might find some ideas that lead to improvements in your security program.

Why Market Your Program

As we begin to look at the merits and benefits of marketing your security awareness program to your employees, I would like to start with a couple of questions:

Do you view your employees as being a part of your security stack?

Do your employees understand their role in the protection of the organization's digital assets?

Do you engage with your employees to promote the security awareness program?

I ask these questions because so often we hear the phrase "employees are our last line of defense," but do they know that? Secondly, are they adequately trained and understand what they are defending?

Employees are a tremendous resource available to security operations in protecting the organization. Still, we need to prepare them adequately for this role, clearly explain their responsibilities, and provide the necessary training so they can effectively help defend the organization from attack. This is where marketing your security awareness program comes into play. We can use the same techniques marketers use to persuade people to purchase a specific product to also promote the employees' role in our cyber defenses and the techniques necessary to be an active part of those defenses.

Enhancing The Security Stack

Examining the current cybersecurity stack of most organizations reveals tools for edge defense, email analysis, endpoint protection, data loss prevention, and a range of other applications, all designed to protect digital assets. We deploy these tools to create a layered defense, as no single tool can sufficiently protect organizations from attack. We refer to this inability to provide full coverage as "gaps." The concept behind the layered approach is to cover gaps in our protection by using multiple products to create a fortified defense. Each product in our security stack serves a defensive purpose and fills gaps in other tools that are not effective defenses against specific attack vectors.

A second problem with the tools in our security stack is that none of them are infallible at defending against the specific attack they are designed to prevent. This might be due to a missing patch, incorrect configuration, or an advancement in our adversaries' attack methods. These tools are all based on technology, and even with advancements in technology, such as AI, they are still susceptible to allowing an attack through or misidentifying a "false negative". What if we had that one tool in our defensive arsenal that didn't rely on technology and had minimal gaps in coverage?

Figure 1.1: *Layered Security*

If we successfully integrate marketing and employee engagement into our security awareness program, they can fill a critical layer of our defense that does not rely on technology, thereby minimizing gaps in our defensive capabilities. The layered security image (**Figure 1.1**) depicts our technological controls and their gaps, and illustrates how a strong human element can form a comprehensive security control around our organization's digital assets. The technological components of the security stack are essential to the organization's defense and cannot be solely replaced by a strong security awareness program. However, a strong human element in our security stack can significantly impact our ability to prevent the penetration of our network. Unlike our technological security controls, humans can assess a potential threat without relying solely on algorithms to make the judgment. This ability to question a threat is based on the sense that something does not feel right, which separates the human defensive element. We need to leverage and capitalize on this characteristic to further strengthen our defenses. This skill must be expanded upon and reinforced through the security awareness program.

The importance of including humans in our security stack is not lost on the attacker's propensity for gaining access to an organization's digital assets by deceiving the human element. Engaging with employees to prepare them for the variety of attack methods they will face helps solidify their readiness to defend the organization adequately. If employees are unaware of the potential attack methods, it puts the organization's data at a much greater risk. Everything that goes into protecting the organization is based on assessing the risk it poses to the digital assets, and this is true for the human element as well.

Evolution of Risk Management

Risk management is one of those terms you hear frequently when discussing cybersecurity and the protection of digital assets. We have evolved to continually evaluate risk at many levels to determine if there is an actionable threat, but this was not always the case. When I began my information technology career in the 1990s, technology was not deeply embedded in business processes, so IT risk was often excluded and not considered in the organization's business risk management. As business operations became increasingly dependent on technology for daily operations, it became apparent that IT risk directly impacts the overall business risk management. Today, it is almost impossible to run a business that does not rely on some form of technology. This reliance on technology and the rapid exploitation of vulnerabilities has created a need for continuous evaluation of potential risks to prioritize mitigation. As part of this ongoing evaluation, the risk posed by our employees must also be considered to address potential threats effectively.

Human Risk Management

Human risk management underwent a similar evolution, as the human element increasingly became the primary target for initiating a cyberattack. Acknowledging that attackers were commonly targeting human behavior, we began developing tools and security awareness programs to train our employees to identify and avoid potentially malicious content. The adoption of the principles demonstrated through the security awareness program led to the establishment of a cybersecurity culture within an organization. The cybersecurity culture is an extension of the overall business culture, establishing the norms and expected behavior for employees to conduct business activities securely. The evolution continued with the development and advocacy of Human Risk Management.

Human Risk Management establishes the foundational aspects that our employees are part of our security stack. Just like other controls in our security stack, each control has inherent risks. The risks of our technological security controls are evaluated based on the information being protected and the functions the device provides to protect against potential attacks. Ideally, we would like to assess the risk each individual in our organization poses. Human Risk Management enhances our ability to evaluate individual risk based on data such as job title, propensity to interact with training emails, and completion of required training. Technologies such as AI enable us to analyze vast amounts of data about employees and to coordinate the application of safeguards, security awareness training, and simulated phishing emails to minimize the risk posed by individuals.

Updating Human Firmware

I find some irony when speaking with colleagues about the amount of training or simulations they expose their employees to. It is not uncommon to learn that they send simulated phishing emails once a month and conduct training annually to meet a requirement. When I ask how often they patch software or appliances, the frequency shifts to weekly or even daily, depending on the severity of the threat. If your employees are the most attacked part of your security stack, why not dedicate the same amount of time to patching the human element?

Managing human risk should be a key component of the organization's security posture. The individuals who comprise your organization are a key component of your security stack and play a vital role in protecting the organization's digital assets. Like other devices in your security stack, the firmware in the human element needs to be continually updated, just like other security technologies. The human element is constantly exposed to highly dynamic and sophisticated attacks. If you choose to neglect or fail to maintain the human firmware, you miss out on a significant security control and could put your organization at serious risk.

The challenge with updating the human firmware is that individuals learn differently and have different motivations. Imagine having to update the protection software on each workstation individually because of slight variations that prevented a mass update. This would present a formidable task, which is what we face when we need to update our human firmware. This is where marketing techniques are highly beneficial for motivating and persuading the human element to move in a common direction. The common direction we are moving towards is the adoption of our organization's cybersecurity culture. We can use marketing tactics to integrate our human workforce into target groups, enabling us to present tailored messages and motivations that drive a mass movement toward the desired behavior of our security program, thereby fostering a strong cybersecurity culture.

Changing Security Culture

The process of creating or changing cybersecurity culture within an organization is a daunting, multifaceted task that is dynamic and must be carefully considered to make significant changes. This book is not intended to serve as a how-to guide for creating a positive cybersecurity culture, as every organization is unique and has a distinct human risk profile that must be considered when implementing behavioral changes. The concepts in this book are intended to offer thought-provoking ideas and present new perspectives on your cybersecurity program, cybersecurity awareness training, and overall approach to engaging with employees, thereby cultivating a strong cybersecurity culture that aligns with the organization's overall culture.

Just as each human is unique, so too is each organization's cybersecurity culture. It is influenced by many factors, such as the size of the organization, the number of locations, locations in other regions or countries, support for work-from-home employees, regulatory requirements, and the organization's leadership's overall commitment to protecting digital assets or appetite for risk. Cultural considerations for an organization operating from a single location with fifty employees will be drastically different from those of a large multinational organization that must also account for local cultural differences within its workforce. Identifying these factors and understanding their role in shaping the current cybersecurity culture is where this journey begins. Understanding these factors will help you identify ways to motivate and engage your employees, encouraging them to actively participate and promote your cybersecurity culture.

Changing an organization's culture is one of the most challenging tasks you can undertake, and this challenge should be approached as a long-term commitment that will not occur effortlessly or instantaneously. As you initiate this transformation, remember that humans, by nature, tend to fall into routines or patterns of behavior. Therefore, changing their behavior to align with your desired cybersecurity culture requires your employees to adopt and embrace the new routine you are promoting. The foundation for changing these behaviors begins with developing a well-rounded cyber-security awareness training program that reinforces the security principles of the desired cybersecurity culture.

Cultivating this new cybersecurity culture begins with a cybersecurity awareness training program that promotes and emphasizes the acceptable practices you want your employees to use in their daily routines. Remember, humans tend to be resistant to change, so the first obstacle to creating the desired cybersecurity culture is overcoming employee behavior that is not aligned with the organization's expected behavior. Changing human behavior must be accomplished through repetition and encouragement; therefore, your security awareness program should be designed to consistently encourage employees to utilize and demonstrate actions that align with the organization's expectations.

2

Why Develop a Strong Cybersecurity Culture?

Before developing your cybersecurity awareness program to enhance your organization's culture, it is essential to understand the rationale behind this endeavor. This will often require justifying the time and cost involved in developing and implementing the program.

Humans are one of the most attacked resources you have in your environment, and they can either be a liability or an asset to your cybersecurity program. The Global Risks Report 2022, published by the World Economic Forum, stated that 95% of cybersecurity incidents are attributable to human error. The human factor in your cybersecurity defenses can sometimes be the first line of defense protecting your organization, or the last line of defense preventing a data breach. According to Proofpoint's 2024 State of the Phish report, 59% of employees were either unsure or did not believe they had a role in the organization's security. Understanding the human role in establishing your cybersecurity defenses is critical, and this role must be explained and reinforced repeatedly to your employees to foster a culture in which employees are part of the defensive solution.

As part of your cybersecurity stack, there is a cost in developing a strong cybersecurity culture. There will be costs associated with the training program itself, including production costs due to employees' time spent on training, as well as ancillary expenses. You must understand the cost of the program and be prepared to justify those costs as you do for other elements of your organization's cybersecurity defenses.

Human Element

Examining human risk management, explaining the human role in data breaches, and developing methods for presenting this information to management are crucial to securing the backing and funding required to create an effective security awareness program. Management must understand that investing more financial resources into the cybersecurity awareness program can help minimize human risk and offset potentially much larger costs associated with a data breach for a fraction of the cost.

Most data breaches begin with human involvement, typically via a phishing email, compromised credentials, or another method that tricks the user into initiating a malicious action. According to the Verizon Data Breach Investigations Report (DBIR) for 2025, sixty percent (60%) of data breaches had a human element. Ironically, the human factor has the highest propensity for a data breach, yet it is often the line item in the cybersecurity defense budget that receives the least funding. Most organizations allocate a significant portion of their cybersecurity budget to technological controls that can be circumvented by human actions, ultimately leading to data breaches. It is essential to demonstrate to management that investing in the human element is just as important, if not more important, than investing in more costly technology-driven security defenses. Malicious actors continually play a cat-and-mouse game, exploiting weaknesses in technological controls that often require a human element to perpetrate the attack. Adequately funding the training of the human element can serve as reinforcement and an additional layer of security to protect against these exploits.

Cost of a Data Breach

Developing the programs to support the transformation of your cybersecurity culture requires funding, and to get this funding, you will likely need to submit budget requests for management approval. It can be challenging to calculate the benefit in risk reduction from technological security controls; therefore, calculating the cost-benefits of an investment in an organization's cybersecurity program can be even more daunting. Most organizations allocate minimal financial resources to their security awareness programs and employee training, yet employees remain a primary cause of data breaches. It is imperative that you highlight how increasing spending on your security awareness program can dramatically reduce the likelihood and cost of a data breach by making your cybersecurity culture part of your security stack.

How do you put a cost on investing to develop your cybersecurity culture to be an active part of your cybersecurity defenses? Start by analyzing the costs associated with a data breach and the initial attack vectors most commonly used to perpetrate cyberattacks. As you begin researching information about data breaches, you will find a common theme: a large portion of cyberattacks start with a human element. The following table (**Table 2.1**) is from IBM's Cost of a Data Breach Report 2025 and lists nine categories of data breach initial attack vectors from most prevalent to least prevalent. This list provides ample evidence of the significant role the human element plays in data breaches, as four of the nine initial attack vectors can be directly attributed to human involvement.

Table 2.1: *IBM's Cost of a Data Breach Initial Attack Vectors*

Attack Vector	USD Millions	Human Element
Phishing	4.80	Yes
Third-party vendor and supply chain compromise	4.91	Maybe
Denial-of-service attacks,	4.41	Maybe
Insider error	3.62	Yes
Compromised credentials	4.67	Yes
Malicious insider	4.92	Yes
Vulnerability exploitation	4.24	Maybe
Physical theft or security issue	4.07	Maybe
System error	3.61	No

A strong argument could be made that four of the remaining initial attack vector categories likely could be attributed to human behavior. Based on the propensity for involvement and the dollar volume, it becomes evident that the human element plays a significant role in executing cyberattacks.

We have established that the human element is a considerable factor in cyberattacks and data breaches; now, we need to identify the costs associated with data breaches. This will further justify the request for funding to develop our cybersecurity awareness program and culture. According to IBM's Cost of a Data Breach Report, the global average cost of a data breach was $4.44 million (US dollars), while the average cost of a data breach in the United States was $10.22 million (US dollars). These figures represent only the average hard costs associated with the data breach, but there will be numerous other costs related to this event that may take years to fully recover from. Cybersecurity insurance should cover part of the costs, but the claim will likely lead to higher premiums in future years. This increase in premiums may be accompanied by a reduction in coverage, potentially leading to a point where the cost of cybersecurity insurance exceeds the benefits it provides. Large organizations should have the capacity to absorb the financial losses related to a data breach, but small and mid-size organizations may face a significant detrimental impact that may threaten their existence.

The IBM report provides a wealth of information and statistics to support the request to fund your cybersecurity awareness program. The dollar amounts involved in data breaches can be significant, but a key takeaway from this information is that human actions are often at the root of the cause. This supports the idea that properly trained employees, committed to the security culture, should be able to take the appropriate actions to prevent the introduction of malicious content into the organization's network, which could lead to a costly data breach.

Brand Reputation Damage

We have already established from IBM's Cost of a Data Breach that the actual cost of a data breach is likely several million dollars. Still, there are additional hidden costs, such as damage to brand reputation from a data breach, that extend well beyond the initial recovery costs.

It takes many years to build a strong brand that your customers know and trust, but it can all be lost in an instant with a single mouse click. That simple mouse click may unleash a cyberattack that damages your network, bringing your business to a halt, causing possibly weeks of lost productivity, significant financial losses, and that is just the short-term costs. The September 2025 data breach at Jaguar Land Rover illustrates the short-term devastation of a data breach, as this incident shut down production for five weeks, resulting in estimated losses in excess of $2.5 billion (US dollars). The long-term costs of the cyber incident may be felt for many years after the initial event, and the organization may not ever fully recover the value of its brand.

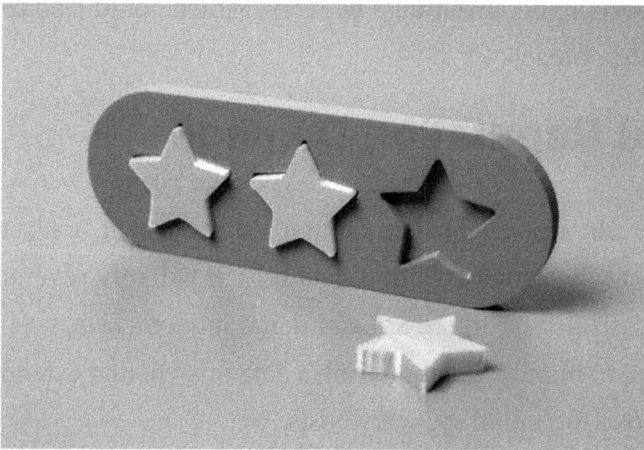

Figure 2.1: Brand reputation damage can take years to recover from
Source: Andrii Yalanskyi/Shutterstock.com

Brand reputation damage refers to how an organization's or product's image is negatively affected by an event that alters the public's perception of its value or trust. This can be particularly significant for organizations operating in industries where consumer trust is paramount, such as financial services. If your customers lose trust due to a security event, it will often take years and significant investment for the organization to regain the public's trust.

Potential impacts of brand reputation damage:
- Loss of customer or business partner trust
- Inability to maintain or grow market share
- Reduced investor confidence
- Difficulty recruiting or retaining employees
- Long-term stigma

The cost of overcoming brand reputation damage typically exceeds the actual cost of the data breach, and the fallout lasts significantly longer. Reviewing a couple of recent data breaches (e.g., SolarWinds), it doesn't take long to find a tremendous amount of information about the difficulties and expense organizations experience when trying to recover from damage to their brand reputation.

Brand reputation damage is a real cost and energy drain to an organization. It is often easy to put a dollar value on the "hard" costs of a data breach, as those reflect tangible amounts such as investigation costs, remediation, legal fees, regulatory fines, and lawsuits. The "soft" costs, such as loss of consumer trust or loyalty, are much harder to put a dollar value on; however, these "soft" costs carry a long-term stigma that often has a more significant impact than the initial data breach. The damage to an organization's brand will likely not put it out of business, but growth will potentially be slowed as resources and marketing efforts will need to be redirected to restoring the brand's image and retaining the current market share.

Missing the Mark

Target sustained a substantial data breach in December 2013 that exposed over 40 million payment card numbers and personal information for approximately 70 million customers. The attackers were able to infiltrate Target's point-of-sale (POS) system through a connection with a third-party vendor. The effect of the breach was imme-diate, occurring in December, a historically significant period when retailers typically account for a large portion of their annual sales. Target's sales were substantially down compared to the previous year, and the stock price declined sharply as a result of this data breach.

Figure 2.2: Target 2014 data breach
Source: bluestork/Shutterstock.com

The breach significantly damaged Target's brand by eroding the trust consumers had in the retailer to protect their sensitive information. This loss of trust persisted for several years, as Target experienced lower sales and revenue. The years of building trust with loyal customers were lost in an instant, and the reputational damage caused by the breach took Target years to fully repair.

Trust, whether between individuals or with an organization such as Target, takes years to develop but can be lost in an instant by a single bad action. It is difficult to put a dollar amount on the cost of fostering the loyalty that existed before this breach, but all the previous advertising, goodwill, and customer development were damaged or lost. In addition to lower post-breach sales, Target had to invest heavily to reconnect with its customer base and reestablish the trust and loyalty that existed before the breach.

References

Target Data Breach Explained: A Case Study
https://www.breachsense.com/blog/target-data-breach/

Data Breaches Cause Loss of Customer Trust
https://www.breachsense.com/blog/data-breach-trust/

Risk Mitigation Strategy

We have previously established that human risk management is the process that we will actively use to reduce the risk presented to the organization by employee action or inaction. When approaching management regarding funding for your security awareness program, I have found that a risk mitigation strategy has worked well. Upper and executive management may not fully understand the technical details involved in a data breach, but they do understand the concept of return on investment (ROI). The previous sections have provided details on the hard and soft costs associated with a data breach. It is your responsibility to present this information in a clear and meaningful way to secure funding for your security awareness program. Rarely is there a true ROI on security investments, so we need to utilize a special formula to calculate the return on security investment (ROSI). Luckily, there is a widely adopted quantitative risk analysis formula developed by The SANS Institute for estimating ROSI. Several calculations are involved in the ROSI. We will begin building our ROSI by performing a standard risk calculation, Annual Loss Expectancy (ALE).

Annual Loss Expectancy

Annual loss expectancy (ALE) is the total annualized loss expected to occur from a particular type of security incident if no action is taken to mitigate the potential risk. ALE is calculated by multiplying the single loss expectancy (SLE) by the annual rate of occurrence (ARO).

$$ALE = SLE \ x \ ARO$$

Single Loss Expectancy (SLE) refers to the monetary loss that would result from a single security incident. When calculating the SLE, you must account for both hard costs and soft costs associated with a security incident. Often, the soft costs will be significantly larger than the hard costs of recovering from the incident.

Annual rate of occurrence (ARO) is the frequency with which a security incident occurs in a given year. For example, if you expect one ransomware attack every three years, the ARO would be:

$$ARO = \frac{1}{3} = 0.33$$

Return on Security Investment (ROSI)

Most investment in security controls is intended to prevent or mitigate losses, so there is no return on investment, as the control will likely never generate a net income. Instead, we must convey the potential loss avoidance as a percentage of the cost of the security control to protect the asset, thereby justifying the investment. This will help validate the expenditure on security controls and prevent spending more money on protection than the asset is worth.

ROSI provides a quantitative calculation of the expected return on investing in a security control to protect an asset. ROSI is calculated by multiplying the ALE by the mitigation ratio minus the cost of the solution, then dividing by the cost of the solution.

$$ROSI = \frac{(ALE \; x \; Mitigation \; Ratio) - Cost \; of \; Solution}{Cost \; of \; Solution}$$

Mitigation Ratio defines the percentage by which the proposed security solution would address the issue or mitigate the risk. From a security awareness training perspective, this is determined by calculating employees' effectiveness in avoiding potential email threats that reach their inboxes. Since the Mitigation Ratio is represented as a percentage, it is calculated by subtracting the current click rate from 100%.

Mitigation Ratio = 100% - Click Rate Percentage

Mitigation Ratio = 100% - 2% = 98%

We use the click rate percentage to calculate the Mitigation Ratio because we aim to demonstrate a risk reduction, based on the effectiveness of our security awareness training program, that improves our employees' ability to avoid potentially malicious emails. We have established that a high percentage of cyberattacks are initiated via email, so the click rate reflects our current risk level, which we are attempting to reduce to avoid potential loss.

POINT OF EMPHASIS: When making a case to management to provide funding for the security awareness program, emphasize that the program is part of a risk mitigation solution to prevent employees from interacting with emails that may contain malicious content. This is an important factor, as technological email security solutions are not fully effective at blocking all potential threats, as attackers can change attack methods that circumvent these controls. This could result in malicious emails reaching employees' inboxes and, if acted upon by employees due to a lack of training, significantly alter the frequency of the ARO. We are justifying the cost of the security awareness program as a risk mitigation measure to prevent employees from interacting with emails that may contain malicious content, because in these circumstances, employees may be the last line of defense against a successful attack.

ROSI Phishing Example

Let's examine how we can utilize ROSI to justify funding our security awareness training program. We will allocate $10.22 million to our SLE, as established by the IBM Cost of a Data Breach Report, which represents the US average cost of a data breach. The cost of the solution will be the amount you want to spend annually on your security awareness program to foster strong employee involvement and solidify your organization's cybersecurity culture. For our example, we will request $15,000 to fund our program, with the expected goal of reducing the click rate on our training emails to 2%. As previously discussed, this would result in a mitigation ratio of 98%.

$SLE = \$10.22M\ (US)$

$ARO = 0.33$

$ALE = \$10.22 * 0.33 = \$3.37$

$Cost\ of\ Solution = \$15,000$

$Mitigation\ Ratio = 98\%$

$$ROSI = \frac{(3,370,000 * 0.98) - 15,000}{15,000} * 100$$

$ROSI = 21,917\%$

Based on these numbers, the request for funding for your program would result in a 21,917% loss avoidance for each dollar invested. This provides a tremendous return on reducing the potential loss. In this example, the $15,000 for your program is likely to be comparatively less expensive than the majority of your technical security controls; however, this investment is in the most vulnerable element of your security stack. This remains consistent with the fact that investment in the human element, although it is often the most vulnerable, is frequently the least funded component of a security program.

Making Your Case

Management will want you to justify funding your cyber-security awareness program. We have reviewed several well-respected reports that indicate the human factor is often the initial attack vector for data breaches. This supports the notion that increasing funding to focus on the human element can help protect the organization from a data breach. The reports have also established the average cost associated with a data breach that we can use to calculate a return on investment. Through accepted risk management formulations, we can demonstrate a significant return on investment in our security program. These factors should help sway management to allocate the requested funding, but there are additional methods you can draw on to solidify support for your program from management.

Your organization's management is driven by optimizing the profitability and financial sustainability of the organization. In addition to hard facts, look for ways to demonstrate the negative impact a data breach may have on the overall business and specific business units. Focusing on how a data breach could affect specific business units will potentially gain the backing of those managers. Start by gathering examples of security incidents that have affected other organizations in your industry and make notes about key facts such as the number of days the entity was down, regulatory fines, lawsuits, and any other financial loss sustained. Use this information to provide hypothetical examples of the impact these events would have on various departments within your organization.

Ask the operations department how the organization would be adversely affected if production were shut down for X number of days?

Ask the finance department how revenue would be adversely affected if orders were not processed for X number of days?

Ask the human resources department how payroll would be handled if it were a manual process due to a cyberattack taking the payroll system offline?

Ask the sales team how the inability to produce products or process orders for X number of days affects our customers, and whether this type of event would drive them to a competitor?

Ask marketing what the cost would be to regain market share that could potentially be lost due to an extended outage and inability to provide our customers with products?

Ask how the organization would function if we had to revert to manual processes versus how the business operates using technology? What would this do to employee morale?

The above scenarios frame the request for funding the security awareness program from a business perspective, highlighting the need to maintain steady business operations. The scenarios likely have a substantial cost that would greatly overshadow the amount you are requesting for your security awareness program. Highlight how funding the program will increase employee awareness, reducing the likelihood that the most vulnerable element in your security stack will initiate a data breach, potentially affecting the organization's viability and integrity.

Finicky Felines

I will be upfront that I have never owned a cat, so it was news to me that simple changes, such as a new type of kitty litter, can be very disruptive to our feline friends. I was unaware of the species' idiosyncrasies and their reluctance to change until I spoke with friends who are cat owners and conducted some research. Through discussions with my friends and some internet research, I soon learned that cats are creatures of habit and that simple changes, such as a different type of kitty litter, might lead some felines to boycott the litter box. So, what does kitty litter have to do with gaining support from management for your security awareness program?

On August 14, 2023, Clorox, which manufactures several popular brands of kitty litter, including Fresh Step and Scoop Away, suffered a substantial cyberattack that took its IT systems offline, affecting operations across the organization. The data breach interrupted production of their kitty litter products. Although Clorox implemented its business continuity plan, production remained offline for several weeks, and full production was not restored for over a month. This lengthy interruption in production, combined with the implementation of a manual order-processing system as part of the business continuity plan, led to product shortages, forcing cat owners to switch to competing brands reluctantly.

Figure 2.3: *Cat checking out the litter box*
Source: Sari Oneil/Shutterstock.com

As we have established, felines are creatures of habit, and changes to necessities such as kitty litter are not always welcome. The Clorox kitty litter shortage compelled pet owners to rapidly transition their pets to a competing product, while dealing with their pets' stress and anxiety as they adapted to the new product. Due to cats' finicky nature and the stress that change can cause, owners were hesitant to transition back to the Clorox kitty litter brands once production was fully restored. The data breach had effectively conceded a portion of Clorox's kitty litter market share to its competitors. If Clorox hopes to regain the lost market share, it will require substantial marketing and promotions to woo former customers back to its products.

This provides a strong example you can present to management to justify funding for your security awareness program. A large portion of breaches begin with the human element interacting with malicious content that could potentially bring the entire business to a halt. Emphasize that investing additional funds in your security awareness program can help prepare your employees to avoid these types of negative situations and the ongoing consequences that often follow.

References

How to Change Brands of Litter
https://petshun.com/article/can-changing-cat-litter-brand-make-cat-sick

Do cats care if you change litter brand?
https://enviroliteracy.org/do-cats-care-if-you-change-litter-brand/

A Bleach Breach: Timeline of the Clorox Cyberattack
https://nationalcioreview.com/articles-insights/news/a-bleach-breach-timeline-of-the-clorox-cyberattack/

Clorox Fresh Step Struggles to Win Back Cat Litter Shoppers
– Bloomberg
https://www.bloomberg.com/news/articles/2024-09-13/clorox-struggles-to-win-back-cat-owners-to-fresh-step-litter-after-shortage?embedded-checkout=true

Summary

An organization's security consists of many layers, and the human element is part of that stack. Unfortunately, the human element is a primary target for malicious actors and often contributes to the initial entry point for an attack. Human risk is like other risks, which can be mitigated to an acceptable level. Understanding the high propensity for breaches driven by human risk, you can build a case to secure funding to reduce the risk.

Additional References

The Global Risks Report 2022 – The World Economic Forum
https://www3.weforum.org/docs/WEF_The_Global_Risks_Report_2022.pdf

2024 State of the Phish – ProofPoint
https://www.proofpoint.com/sites/default/files/threat-reports/pfpt-us-tr-state-of-the-phish-2024.pdf

IBM Cost of a Data Breach Report
https://www.ibm.com/reports/data-breach

2025 Data Breach Investigations Report | Verizon
https://www.verizon.com/business/resources/reports/dbir/

Brand Reputation Damage

Can SolarWinds survive? For breached companies it's a long, painful road to restoring trust
https://www.scworld.com/news/can-solarwinds-survive-for-breached-companies-its-a-long-painful-road-to-restoring-trust

How JLR's Category 3 Cyber Attack Caused Production Shutdown *https://cybermagazine.com/news/jlr-cyber-breach-financial-disaster*

Risk Mitigation

Quantitative Risk Analysis Step-By-Step
https://www.sans.org/white-papers/849/

The One Equation You Need to Calculate Risk-Reduction ROI
https://www.cisecurity.org/insights/blog/the-one-equation-you-need-to-calculate-risk-reduction-roi

Calculating Cybersecurity ROSI
https://www.corsicatech.com/blog/cybersecurity-roi-rosi-calculator/

How to Calculate Return on Security Investment
https://blog.netwrix.com/2018/08/07/how-to-calculate-return-on-security-investment/

Unlocking the Value of Cybersecurity: Calculating ROSI (Return on Security Investment)
https://www.linkedin.com/pulse/unlocking-value-cybersecurity-calculating-rosi-return-pramod-kuksal-5sznc/

3

Marketing Fundamentals

(Having a marketing degree) I often joke that a marketing degree is a psychology degree with a focus on business principles. This makes more sense when you understand that portions of psychology involve the study of perception, emotion, and motivation, all of which are essential elements in positioning your product to fulfill a customer's need or desire. When marketing your product, you want to create a distinct perception of it that elicits a positive emotion from your customers. This positive emotion fosters a desire for the product, thereby increasing the customer's motivation to purchase it, or, in our case, actively participate in our cyber-security culture. The product you are selling to your employees is security awareness, so you will need to be creative in identifying ways to evoke a positive perception of your program that generates strong motivation for participation.

In this chapter, we will cover foundational marketing concepts integral to developing a marketing approach to expand our cybersecurity culture throughout our workforce. This is not intended to be a comprehensive exploration of marketing topics, as numerous books have been written on this subject. The marketing topics covered provide a basic understanding of key marketing concepts to help you get started in using marketing strategies to promote and build your security awareness program. We will begin by examining the difference between our marketing strategy and our marketing plan. These terms are often used interchangeably, but they are distinctly different in their purposes and equally important to successfully marketing your product.

The journey into marketing our security awareness program begins by familiarizing ourselves with some key concepts that will be instrumental in creating a successful program.

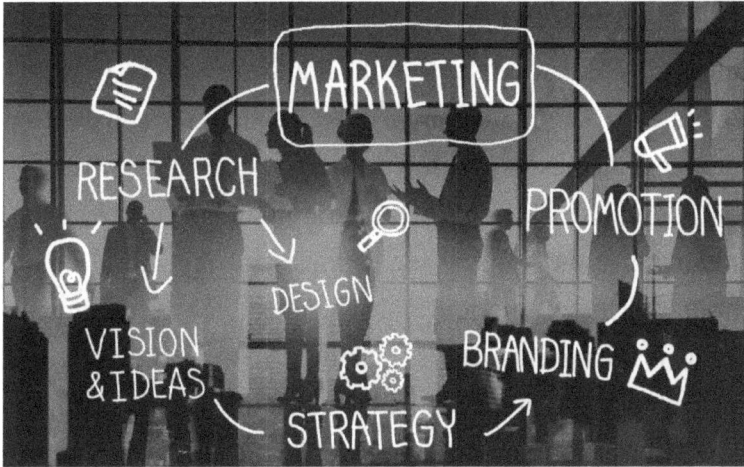

Figure 3.1: *Marketing Concepts*
Source: Rawpixel.com/Shutterstock.com

Branding

Branding is a critical element in developing your product's marketing strategy, but there is much more to branding than just creating a flashy logo or catchy slogan. Branding and brand are different elements that work together to identify and market your product.

Brand is the method of using names, logos, colors, and taglines to make the product identifiable in the market. Creating a successful brand requires more than just creating recognition; it is a way to establish trust and a reputation with your consumers. The brand's reputation is defined by consistent experiences with the product and its associated marketing. These elements of the brand are used by branding or brand marketing to promote brand recognition and develop market strategies to position the brand in the marketplace.

Brand marketing (branding) promotes the brand by highlighting the values, perceptions, and reputation of the product. This promotion is designed to elevate the brand's identity among consumers, making it more identifiable and differentiating it in the market. When implemented correctly, consumers will immediately associate the product or service offered with the brand's recognition. The brands shown in **Figure 3.2** are well recognized, and most individuals immediately identify the products or services associated with these brands. This recognition is likely associated with other factors such as trust, reliability, and prestige, further demonstrating the power of a brand's perception with consumers.

When consumers have a favorable view and trust a product's reputation, it can establish brand loyalty with the consumer, often eliciting emotional connections. Generating positive emotional responses makes it much easier to increase your market share through well-defined marketing strategies.

Figure 3.2: *Familiar brands*
Source: Bashigo/Shutterstock.com

Market Share

Increasing market share is often a key goal of a marketing plan and overall marketing strategy, but what exactly is market share? Market share is a marketing metric that provides information about a product or brand's control over a market relative to similar products in the marketplace. A simplified approach to understanding this is to consider the product or brand's popularity among consumers.

Market share is expressed as a percentage based on the product's sales divided by the total sales volume of the market.

$$Market\ Share = \frac{Total\ Sales}{Total\ Market\ Sales} \times 100$$

Maximizing and increasing market share is commonly a driving factor in most marketing strategies and plans. Market share is increased by taking customers away from competing products, which are often an indicator of successful marketing strategies and plans. Increasing market share is achieved through marketing campaigns that highlight and emphasize the benefits of your product in comparison to the competition. This may include demonstrating superior alignment of your product with the customer's values and ability to meet expectations. As you increase your market share, you will begin to exert market dominance, thus increasing your brand recognition and customer loyalty.

Analyzing market share continually provides invaluable feedback about market direction and consumer perceptions of your brand in relation to competitors. This analysis encompasses awareness of market trends and potential growth opportunities for your product. The analysis of this data allows marketers to understand the brand's strengths and identify target audiences for potential growth. The information ascertained from this analysis is used to adjust current marketing plans and define the purpose of future marketing plans.

Develop A Marketing Strategy

The marketing strategy is the top level of a nested methodology for effectively marketing your product. Much like Matryoshka (Russian) dolls, there are multiple components inside a larger component that make up the overall marketing program. The marketing strategy is the top-level plan that outlines the long-term goals and objectives for the marketing efforts. Within the marketing strategy, there are marketing plans that provide operational guidance for achieving the goals stated in the marketing strategy. Finally, the marketing plan outlines marketing campaigns that define the promotions and customer interactions designed to influence and motivate behavior, ultimately achieving the overall marketing goals.

The marketing strategy provides a high-level blueprint for the long-term direction of our marketing efforts aligned with the specified business objectives. This blueprint effectively outlines the purpose, plan, target audience, and objectives for promoting your product to achieve the desired results. The marketing strategy does not outline specific actions for achieving the stated goals, but it does provide a vision to align marketing efforts in pursuit of those goals. This includes direction, orchestration, and guidance to maintain consistency in branding and messaging alignment throughout all marketing activities.

Defining your marketing strategy is a crucial first step to ensure that your marketing aligns with the organization's expectations. Let's examine the components that make up an effective marketing strategy.

Purpose or Long-Term Goals

The marketing strategy defines your "why" or the purpose of your overall marketing efforts. What are the long-term goals for your marketing efforts? Often, the purpose of your marketing is to achieve a business objective, such as increasing brand awareness or market share. Clearly defining the purpose of your marketing strategy is crucial to ensure the derived marketing plans align and contribute to the stated business objectives and goals.

Audience

Who are we marketing to?

The marketing strategy identifies the target audience to whom we are marketing. This is often defined as demographics, which identify individuals in the market based on characteristics such as age or location. Understanding the demographics of your audience is crucial to your marketing strategy, as it enables you to create marketing plans tailored to effectively reach your target audience.

Generational Demographics

Generational marketing focuses on segmenting the target audience based on age-related demographics. These demographics refer to specific age boundaries (see **Table 3.1**), defined by milestone historical events and influences that shaped the typical traits of each generation. The events and technological advances that shaped the generation influence brand loyalty and preferred marketing channels. Understanding these traits and how they shape the generation's behaviors provides valuable marketing insight used to develop marketing strategies specifically designed to approach and influence the different generations. While there are commonalities among the generations, it can be challenging to craft marketing messages that encompass all of them.

Table 3.1: *Generations*

Generation	Birth Year
Silent Generation	Prior to 1945
Baby Boomers	1946 – 1964
Generation X	1965 – 1980
Millennials	1981 – 1996
Generation Z	1997 – 2010
Generation Alpha	2010 - 2024

Different Motivators - Cultural Differences

Understanding cultural differences when creating marketing strategies is crucial, as different societal groups perceive marketing messages differently due to their unique beliefs, customs, and cultural norms. Cultural differences may exist between countries or within regions of a country. It is crucial to acknowledge these differences to create effective marketing strategies that resonate with the target culture. Tailor your marketing strategy and messaging to promote your product effectively.

Unique Value Proposition

The marketing strategy should include a value proposition that defines the unique features of your product or service and how those features add value for your customers. The value proposition conveys the benefits customers can expect from using your product in comparison to competing products. This essentially defines the value of your product's distinguishing features, explaining why these features set your product apart from the competition and why customers should choose to purchase it. The creation of a powerful value proposition supports the establishment of a brand identity that can lead to increased consumer loyalty.

Brand Positioning

We previously discussed branding and its impact on our product. The marketing strategy now must define how we will position our brand to stand out in the market. This is where we develop our plans for differentiating our product from competitors in the eyes of our target audience. As with familiar brands, we aim for our marketing strategy to create a memorable, positive impression of our brand that meets the target audience's desires and expectations. If we can effectively position our product in the minds of the target audience, they will associate our product with the means to solve their problem or fulfill their desire.

Strategic Goals

The Marketing Strategy should define clear marketing objectives that the organization hopes to achieve as part of the marketing initiatives. These marketing objectives should be specific, measurable, achievable, relevant, and time-bound (SMART). Key performance indicators (KPIs) track the performance of marketing initiatives in relation to marketing objectives and serve as the basis for evaluating whether these initiatives are producing the desired results. Based on the analysis of the KPIs, program managers may adjust the marketing initiatives to meet the established marketing objectives.

Develop A Marketing Plan

The marketing strategy provided a high-level blueprint for the long-term direction of our marketing efforts, aligning with the established business objectives. The marketing plan breaks down these blueprints into smaller tactical steps to achieve the stated goals of the marketing strategy. The marketing plan is based on operational objectives that collectively contribute to achieving the overall goal stated in the marketing strategy. This is essentially the "how" we will reach the desired business objectives. When developing a marketing plan, the efforts and results must align with and contribute to the objectives defined in the marketing strategy. The marketing plan defines short-term goals, marketing tactics, target segment, budget, and metrics to measure the success of the plan and its contribution to the marketing strategy.

View the marketing strategy as an expedition, and the marketing plans are the smaller journeys taken to reach the desired destination. Each smaller journey will have its own goals, objectives, and challenges that must be overcome to complete the expedition. Each of the smaller journeys is a crucial step toward achieving the desired outcome.

Executive Summary

The executive summary provides a concise overview of the marketing plan's objectives. This will include the underlying theme and the target audience. The summary will also discuss how this plan contributes to the overall marketing strategy.

Marketing Goals

The overarching marketing goals were defined in our marketing strategy. What are the goals of the marketing plan, and how does it contribute to the overall marketing strategy? The marketing plan should move us closer to the larger marketing objectives, so the goals of the marketing plan will also focus on a segment of the larger goal.

Market Analysis

Analyzing your program's strengths and weaknesses is a crucial first step in developing your marketing plan. The marketing plan is designed as a small step towards achieving the overall, larger objective of the marketing strategy, which is to influence your security culture. Understanding the current state of your security culture enables you to set realistic and achievable goals for the marketing plan you are developing.

Analyzing your current metrics additionally may provide insight into ways to maximize your marketing efforts. As your security culture matures, making changes will become more challenging because the room for improvement is smaller. Use your metrics to determine which areas of the security culture provide the most benefit to the organization's overall security. Refine this analysis further by reviewing which of these areas your marketing efforts can affect with substantial change, with the least amount of effort. Just as with risk analysis, determine where to allocate your resources (time and budget) based on your ability to effectively mitigate the most likely risk facing the organization.

Campaigns and Tactics

The marketing plan has defined the operational objectives that will be achieved through the marketing campaigns. Depending on the marketing tactics, the marketing plan may utilize several different marketing campaigns to achieve operational goals. Marketing campaigns put the marketing plan into action by capturing the target audience's attention through coordinated messages across various channels. The messaging should inform, engage, and influence the viewer to alter their behavior to the desired response or perception of the brand.

Marketing campaigns for our security awareness program will not have all the channels available to traditional marketers. These campaigns will be limited to internal communications, which may include the following channels:

- Company Intranet

- Email marketing (newsletters)
- Communication platforms (Slack, Teams)
- Printed materials (brochures, posters, stickers, swag)
- Public relations

Presenting the message consistently across various channels is crucial to maximizing the campaign's impact. Consistency with your campaign's messaging reinforces your brand and your product with the consumer, which in this case is your employee. Across all our marketing efforts, whether the desired effect is to improve the brand recognition of your security training program or to target a specific area for improvement, the messaging should consistently build on and reinforce our organization's cybersecurity culture.

Target Audience

Who is your marketing plan's message intended for?

How will your marketing campaign's message influence your audience?

The marketing strategy defines the larger, encompassing audience our marketing efforts aim to influence. We touched on demographics and the importance of understanding the makeup and characteristics of our target audience, as this directly affects how we create and deliver the message of our marketing plan.

Depending on the size of your organization and geographical operations, the demographics of your workforce may be very diverse and unique. Unlike traditional marketing, where the holistic audience may not share any common characteristics, our audience does share at least one characteristic: they are all part of the organization's security culture. This unifying characteristic creates opportunities to craft marketing plans that raise brand awareness of your security program and underlying cybersecurity culture.

Creating marketing plans and campaigns to reach audiences with diverse demographics can be a challenging task. The diversity of the larger demographics may reduce the effectiveness of our messaging or make it less relatable to a portion of the broader audience. Understanding how the differences in our larger audience affect our marketing plan is critical in developing marketing campaigns that reach the desired segment of our audience to achieve the desired results. In marketing terms, this is referred to as "market segmentation."

Market Segmentation

Market segmentation is a method for dividing the larger audience defined in the marketing strategy into smaller groups based on common demographic characteristics and interests. These smaller groups or segments allow marketers to craft custom marketing campaigns based on their shared attributes. These attributes may be based on common interests, preferred communication channels, pop culture events, or geographical location. The basis for creating these market segments is rooted in the belief that members of these groups will respond similarly to marketing efforts that target their shared characteristics. The marketing plans and related campaigns often target these market segments rather than the larger overall market because it is a more effective approach to deliver the message to achieve the desired results.

Budget

A budget will be required to execute effective marketing campaigns as outlined in the marketing plan. Some campaigns may be conducted with no budget, while others may have a budget of $100, and others may involve several thousand dollars. We previously detailed using ROSI to demonstrate how a modest budget for security awareness training can go a long way in reducing one of the most significant risks to the organization: the human risk. We must manage the allocated budget to maximize our influence to reduce the human risk. The key is to optimize your budget throughout the year to make the most significant impact on your employees and your cybersecurity culture.

Develop a calendar outlining your marketing plans for the year, if possible. Identify which campaigns may require budget allocations. Some campaigns may be very short and require minimal funding, while others may span an extended period, involving various interactions and promotions with your employees, and necessitate larger allocations. As part of your yearly planning, develop campaign budgets from your overall budget for each campaign. This will help you manage your spending to ensure each campaign is effective and funded appropriately.

Campaign items that may require budget allocations:

- Security awareness training platform
- Printed materials
- SWAG (promotional items, t-shirts, mugs, etc.)
- Meals for lunch and learns
- Employee Engagements

Manage your budgets to ensure each campaign is appropriately funded, thereby maximizing its impact and positively influencing the desired employee behavior. The campaigns must deliver a return based on the ROSI you presented.

Key Performance Indicators

The marketing strategy defines the long-term goals and key indicators we aim to achieve through the marketing of our security awareness program and the adoption of the organization's cybersecurity culture. The KPIs for the marketing plan may track a specific aspect tied to the plan, but often the KPIs for marketing plans are intermediate goals to achieving the overall strategic objectives.

Common key performance indicators:

- Decrease simulated phishing click rate
- Increase security awareness training completion
- Increase reporting rate
- Reduce code vulnerabilities

Determine which goals defined in the marketing strategy will be potentially affected by the marketing plan. Develop realistic intermediate goals based on the strategic goal outlined in the marketing plan and its related campaigns. Set marketing plan goals that are realistic and achievable but still represent a solid challenge and growth towards the overall strategic goal. The marketing plan KPIs measure how well the campaign contributes to the strategic goals and overall success of our program.

Bringing A New Product to Market

We will discuss some key considerations for introducing a new product to the market. Whether you are inheriting a well-run security awareness program, revamping an existing program, or starting a program from scratch, the concept of bringing a new product market applies. Your organization likely has a cycle of employee turnover, so you continually get new people who are unaware of the product you offer. You must introduce your product and its benefits to these new employees to attract them to your market share, thereby becoming active participants in your organization's cybersecurity culture.

Bringing a new product to market is never an easy task, and many new products fail to succeed. One reason new products fail is that they are unknown to the consumer, so there is no history or understanding of the product's benefits. The latest product may offer many unique and desirable features, but consumers may be hesitant to purchase it due to uncertainty about whether it will fulfill their needs. Have you gone to buy a product for the first time in an online store, but did not complete the purchase because there were no reviews for the product? Compare this to the number of first-time purchases you made when there were numerous reviews discussing the pros and cons of the product. This additional information likely provided the necessary reassurance that the product would meet your requirements.

Another factor new products may face is that they are creating a new market space where there is no established consumer need. Many new products enter the marketplace without an identified consumer need, but because of innovative features or popularity, they become desirable, creating a new market.

Building or changing your cybersecurity culture is no different, as you are essentially introducing a new product to your employees. You may have inherited an existing program that may be underperforming, so you will need to market the "new features" of your security awareness program. Your organization may not currently have a security awareness training program, so you are creating a new product for a market space that does not currently exist. Remember, new products often fail because the message to the consumer does not resonate, creating a lack of desire to purchase the product. You can increase the odds of employee engagement in your security awareness program by identifying ways to establish value and create an eagerness for employee participation.

Envision your desired security culture as a new product you are introducing to your employees. It is your job to create interesting and enticing methods to promote and educate your potential consumers, in this case, your employees, about the benefits of your product. Demonstrating the benefits of your cybersecurity program and personalizing the marketing message can significantly increase participation in your program if you can show how the program will enrich or improve employees' lives.

Summary

Marketing the security awareness program may be something new, so familiarizing yourself with some marketing concepts is a good starting point. Promoting brand identity or increasing market share are two common goals of the long-term marketing strategy. Identify your target audience and understand that there are differences across its segments, so you will need to find commonalities to develop your marketing campaigns. Establish KPIs to measure progress and to determine the effects your marketing is having on improving the cybersecurity culture.

Additional References

Demographics

Generational marketing explained: Everything you need to know
https://www.techtarget.com/whatis/feature/Generational-marketing-explained-Everything-you-need-to-know

Why Brands are Investing in Cross-Cultural Marketing
https://www.weglot.com/blog/cross-cultural-marketing

Brand Positioning

What Brand Positioning Is And Why It's Important For Your Business
https://www.forbes.com/councils/theyec/2022/01/14/what-brand-positioning-is-and-why-its-important-for-your-business/

Marketing Strategy

Marketing Plan vs. Marketing Strategy: Understanding the Essentials

https://creately.com/guides/marketing-plan-vs-marketing-strategy/

Marketing Strategy vs Marketing Plan: What's the Difference?

https://www.lairedigital.com/blog/marketing-plan-vs-marketing-strategy

Marketing Strategy vs Marketing Plan: Understanding the Key Differences

https://www.launchnotes.com/blog/marketing-strategy-vs-marketing-plan-understanding-the-key-differences

Market Segmentation

Market segmentation — definition, types, and examples

https://business.adobe.com/blog/basics/market-segmentation-examples

Conceptualize The Program's Marketing Strategy

We aim to develop a marketing strategy for our security awareness program to foster a stronger security culture within our organization. Improving the security culture is directly related to our employees' ability to consistently demonstrate the desired behavior of analyzing, avoiding, and evading potential threats. The question then becomes: how do we motivate employees to actively want to change their behavior and assimilate into the organization's security culture?

Before marketing your cybersecurity awareness program, it's essential to understand what your product is, how it relates to your employees, and the benefits it offers. A fundamental element of the marketing strategy will be to reach employees on a personal level, demonstrating how adopting the desired secure behavior benefits them personally.

Program's Purpose

We aim to enhance our organizational defenses against cyberattacks by implementing a robust security stack that leverages multilayered defenses. We spend a tremendous amount of our resources on technical controls to protect the organization's assets, but we often neglect one of the largest attack vectors, our employees. A majority of cyberattacks are initiated by exploiting the human element, so it is imperative that we adequately prepare employees for their role in our cybersecurity defenses.

Our employees are a part of our organization that interacts with outside individuals and technologies that can pose significant threats to our operations. If our employees are not adequately trained to contribute to our cybersecurity defenses, they pose a substantial risk to the organization's security and overall operations. Employees may interact with websites or emails that contain malicious content. Employees may circumvent our defenses (Shadow IT) by using unapproved technologies to achieve perceived efficiencies in their work processes. This type of activity presents significant risk to the organization, but it is preventable. We must educate our employees on why this behavior presents a risk to the organization. By explaining why this type of activity is risky, we lay the groundwork for instructing our staff about their role in defending the organization against cyber threats.

Education

Educating your employees and stakeholders about cybersecurity is the cornerstone of your program. Individuals outside the cybersecurity profession are often unaware of the threats that are directly in front of them. Like sharks swimming just below the surface, your employees may be unaware of the danger that is in close proximity. We must market the message of our security awareness program to our customers (our employees) to capture their attention while remaining engaging and informative to enlighten them about the imminent threats they face. If our employees are unaware of the threats they may encounter, how can we expect them to perform the desired actions and be an active part of cybersecurity defenses?

Education is a perpetual activity of your security awareness program. This ongoing endeavor must evolve as technology changes along with the associated threat landscape. As technology advances, attackers will likely discover new ways to exploit the vulnerabilities that emerge. Training must continually inform staff about techniques they can use to identify and avoid threats while using approved technology safely and securely.

Explore innovative approaches to engaging with employees and educating them about cybersecurity best practices. By stimulating your staff's interest in the security awareness program, you will garner their attention for cyber-security topics. As their interest expands, they will become more invested in the security awareness program and become active participants in the organization's cybersecurity culture.

Promote Employees' Role in Cybersecurity Defenses

Your employees are part of your security stack. Depending on how you want to look at it, they are your first line of defense or possibly your last. Either way, employees represent a significant layer of protection for the organization's networks. Security professionals often discuss layers of defense because multiple products protect the network, as no single technology can provide a foolproof defense against all malicious cyber activity. I believe our employees offer one of the strongest layers of defense, with potential for a very narrow gap for penetration, but they must be adequately prepared for this role.

How can we effectively engage our employees to prepare them for their role in protecting the organization's network? Educating employees about cybersecurity threats is a key initial objective. As we enhance our employees' understanding of cyber threats and attack methods, we then promote their role in the organization's cybersecurity defenses. It is essential to recognize and acknowledge the contributions employees make to protecting the organization's resources. This recognition sends a powerful message to the employees about their contributions to protecting the organization. Acknowledging their contributions as part of the organization's security stack makes it personal and creates a sense of responsibility to continue with their efforts.

Program's Target Audience

When developing a target audience using a conventional marketing strategy, we attempt to identify the demographics we hope to persuade through our marketing activities. Recall that demographics is a marketing term used to categorize people by several general factors, such as age and gender, which are utilized by marketing campaigns to target commonalities. The makeup of your organization is a microcosm of the demographic sets typically used in traditional marketing. Due to the smaller representations and limited budgets, you will likely not be able to create large-scale strategies targeting a specific demographic of your workforce. Therefore, you must develop marketing strategies that reach multiple demographics simultaneously. It is essential to understand the demographics of your organization, as this will directly impact how you develop your marketing strategy to maximize the effectiveness of your cybersecurity program's marketing efforts.

Understanding your audience is crucial when shaping the organization's cybersecurity culture, as motivations for different segments of the workforce vary based on age, gender, location, and other distinguishing characteristics. Recognizing the characteristics of the workforce make-up and underlying subsets helps develop marketing strategies to promote, sustain, and elevate the organization's cyber-security culture. Remember, we are marketing cybersecurity to our employees, so understanding the motivations of the different groups is essential to present a message that resonates and fosters participation in the security awareness program.

This cross-generational make-up of your workforce may be one of your most significant challenges when creating a marketing strategy. Understanding the characteristics and driving forces of a generation provides the basis for determining the most influential way to communicate your message to actively promote the organization's cybersecurity awareness program and associated cybersecurity culture. Creating marketing strategies specific to each generational demographic may not be feasible, so you will need to develop marketing strategies that identify common influences and motivations relevant to all generations in your workforce. As you develop the marketing strategy for the security awareness program, understanding the motivations of the different demographics represented in your workforce is imperative because your underlying goal is to create one cohesive cyber-security culture that all members of your organization recognize. A strong corporate security culture, recognized by all demographics within your organization, would be an ideal focal point for your marketing efforts.

Memories That Bond Generations

Developing marketing activities that involve multiple generations can be a challenging task. It is essential to identify a common focal point that holds meaning for all targeted demographics. You must identify something that relates to all members of your organization, from the twenty-year-olds just entering the workforce to the mature individuals nearing retirement. The common focal point may evoke different memories for each group, but it provides a shared point from which to create marketing activities that resonate with all.

One method I have found effective for creating campaigns that resonate with multiple generations of the workforce is through the use of movies. Especially movies that are part of pop culture and remain relevant across several generations. Depending on which direction you want to take, there are a couple of approaches to consider when selecting movies that might be effective for incorporating into a campaign to promote your security awareness program.

Figure 4.1: *Generations bonding over pop culture*
Source: fizkes/Shutterstock.com

Identify popular movies that the younger generations grew up watching and the older generations remember watching with their kids or grandchildren. This will resonate as childhood memories for all generations involved, but from different perspectives. Although the perspectives differ, there is a common focal point that you can use, as the movie is identifiable to all demographics in your organization. An example we have effectively used in a Cybersecurity Awareness Month promotion revolved around the movie Monsters, Inc. This promotion included a "Scare Floor Leader" list that identified all employees who had not clicked on a simulated phishing email within the specified time period. We incorporated other elements from the film into internal security promotions and swag that was well-received by our staff. This movie resonated with all members of our demographics for different reasons, but it allowed us to create common marketing promotions that were relatable to all our employees.

Another way to look for movies is to consider those that older generations made into pop culture phenomena and then passed down to younger generations as they grew up. Again, the film may evoke memories for different reasons, but it creates a common focal point for your marketing efforts. We ran a Cybersecurity Awareness Month promotion based around Ghostbusters. This was fun because it coincided with Cybersecurity Awareness Month, which occurs in October, so it tied into the Halloween aspect as well. We ran numerous campaigns promoting our cybersecurity month activities, including a phishing derby where you were "slimed" if you misidentified an email. We will discuss this further later in the book. Additionally, we created our own caricature of Slimer as a malicious email, which became a fan favorite among our employees. Our Slimer caricature was featured on swag we handed out throughout the month and even made an appearance during the winter holidays, complete with antlers and lights. The Ghostbusters promotion was a highly successful campaign that generated numerous positive interactions between our employees and security operations staff, promoting best practices in cybersecurity.

There are many ways to find relatable content that spans the generations. Look for pop culture events, such as movies, video games, and music videos, to find inspiration. Your workforce is a smaller sample of the larger demographics, so it can be challenging to find that common focal point. Identify a focal point that evokes a positive memory across your demographics and create campaigns that elicit these memories, allowing you to develop activities around them to promote your security awareness program.

Culture Differences

Cultural differences significantly influence marketing strategies and are among the most significant challenges when developing a security awareness program. If you work with a large multinational organization, you will undoubtedly need to account for cultural differences within your workforce; however, local cultural differences can also exist within the same country or region. These differences will influence the materials you include in your security awareness program for different groups and how you develop your supporting marketing efforts. We aim to maximize participation in our security awareness program, so we will tailor the materials and marketing efforts to align with local cultural norms and values.

Segmentation

Similar to cultural differences, your organization may have different departments or groups that operate differently or require specialized training based on their job functions. The members of your software development team may require specialized training in secure coding. The sales department may spend most of their time in the field, so they may require courses involving the use of virtual private networks (VPNs) or public Wi-Fi. This uniqueness does not preclude these groups from being part of the larger, overarching cyber-security culture your organization promotes, but they do have specialized training requirements that must be incorporated into your security awareness program. Identify specialized groups in your organization that may require unique training and development supplemental programs to help members of those groups identify and evade threats that may specifically target them.

Program's Value Proposition

The value proposition is expressed from the perspective of enhancing the organization's cybersecurity culture through the education and engagement of its workforce via a cyber-security awareness program. Defining the value proposition for your cybersecurity awareness program establishes the benefits of your product in relation to the competition. This is where the marketing aspect becomes a little twisted, as we do not have any actual competitors. You must craft your value proposition to convert users who do not routinely exhibit the desired traits of your cybersecurity culture into active participants. Essentially, you are trying to take market share away from your competitors (employees who have not adopted your security practices) by demonstrating the benefits of your product to persuade those individuals to adopt the organization's preferred security practices.

There are two key elements in developing your value proposition for your marketing strategy. These two elements feed off one another and begin with improving the desired behavior of your employees. The marketing efforts related to the overall strategy aim to reinforce users who already adhere to the organization's defined best security practices, while attempting to persuade those who have not adopted these security practices about the merits of the desired security behavior. As your marketing efforts convert users to embrace the desired security practices, this increases the number of users actively participating in your cybersecurity awareness program.

This increased activity leads to the second element of the value proposition, boosting the organization's cyber-security culture. The cybersecurity culture is driven by a collective adoption by the organization's workforce to actively incorporate the desired security behavior as a habitual component of their work routine. A strong cyber-security culture is established when each employee adopts the defined best practices and expects their peers to adhere to these best practices. As more employees demonstrate their dedication and devotion to using the defined security practices as part of their workplace activities, the behavior becomes ingrained in the workforce and is expected by their co-workers. This indoctrination into the organization's defined security practices fosters a stronger cybersecurity culture, thereby making the organization less vulnerable to threats.

Creating The Program's Brand Identity

The basic principle of our plan is to market our security awareness program, influencing the adoption of the organization's desired security behavior and thereby improving the underlying cybersecurity culture. The communications of the security awareness program likely compete for your employees' attention with other internal communications, so we need to create a brand identity (logo) that your employees identify and connect with to make your communications stand out. A product's brand identity is often tied to brand loyalty, trustworthiness, and reputation, which is what we want to convey to our employees about our security awareness program through branding.

I have found that several key considerations are essential when designing a brand identity for your security awareness program. The organization likely has an established brand identity, and the security awareness program's brand identity is an internal extension of that identity. It is essential to maintain consistency with the organizational brand identity by using the same color scheme and fonts whenever possible. This creates a cohesiveness between the organization's brand identity and the security awareness program's brand identity. This cohesiveness is essential because, like the extension of the brand identity, the cybersecurity culture is an extension of the organization's culture.

When creating the logo for the security awareness program, find something that relates to the organization's overall mission or values. If your organization routinely uses a particular word or phrase, try to incorporate that into your security program's brand identity to maintain compatibility with the organization's branding. For example, if the organization is Community Financial Institution, and the word "community" is frequently used in the name of various organizational programs, you may want to develop the following logo to represent your security program's brand.

Community Safe
Protecting Our Community

Figure 4.2: Community Safe example logo

The sample logo for our fictional Community Financial Institution's security awareness program is represented in **Figure 4.2**. The logo is much more than just images and words; it represents your program's brand and purpose. We reviewed the logos of some major brands earlier in this chapter, and upon seeing those logos, they immediately elicit knowledge and feelings about the associated products. When viewed, your logo should immediately remind the viewer of the mission and purpose of your security awareness program.

Let's dissect this fictional logo to understand some of the methodologies I use when creating logos for a security awareness program's brand. The logo must quickly remind the viewer of the brand's mission to secure the organization's digital assets. In this example, we use the universally recognized image of a safe. Safes have long been associated with financial institutions and immediately create visions of securely storing valuable assets. It was noted that the word "community" is often used in relation to other programs within the organization, so by using the symbolism of the safe image with the phrase "Community Safe", we create a base logo that references our security program's purpose: securing Community Financial Institution's data. We reinforce the mission of our security awareness program with the tagline, "Protecting Our Community." The words in this tagline were specifically selected and serve a purpose. The word "Our" is meant to remind the viewer that each employee has a role to play in "Protecting" the assets of "Community" Financial Institution. Logos are intended to be quick reminders about the brand they represent. In this example, we created a simple logo consisting of an image and five words that quickly reinforce the mission of the security awareness program and promote each employee's responsibility to keep Community Financial Institution's data safe.

Meet My Mascot

Mascots are another effective visual element that can be used to supplement your cybersecurity program's branding. Mascots are typically cartoon characters featured in marketing materials and can be created for various events, such as Cybersecurity Awareness Month, holidays, or other special occasions. We previously discussed the success of our "Slimer" mascot that was part of our Ghostbusters Cybersecurity Awareness Month campaign. These characters often take on a life of their own within the program and frequently become fan favorites with your audience.

Devolutions provide a great example of using caricatures to market products (see **Figure 4.3**). Devolutions is an IT company that produces a line of products for password management and centralized remote access management. The Devolutions' Sysadminotaur caricatures are found throughout their website and other marketing materials promoting their products. A key to using caricatures is to be consistent and create characters that are identifiable with the target audience. The Sysadminotaur characters on the Devolutions site all exhibit a distinct persona related to an area of IT management, which ties to their products, making them fun and relatable to the intended audience. Another outstanding aspect of this site is that the supporting artwork is created in a style similar to the caricatures, resulting in a consistent look and feel throughout the site. Caricatures are a fun way to promote your product or brand and often generate an attachment with the intended audience. The Devolution website demonstrates how to effectively utilize caricatures to market its products and brand.

Figure 4.3: *Devolutions Sysadminotaur caricature*

Source: Devolutions Sysadminotaur Image © Devolutions Inc. and Patrick Desilets, used with permission.

Reference

https://www.devolutions.net

Strategic Goals

Strategic goals set the long-term direction and plan to market our security awareness program. This serves as the backbone for all marketing efforts in support of achieving the defined milestone. The definition of success for the program is established in the strategic goals, which typically span three to five years into the future. When establishing strategic goals, it is crucial to set objectives that are realistic and attainable while pushing the program to achieve greater success.

If your security awareness program and associated cybersecurity culture are immature, you may be able to set ambitious strategic goals, as there is likely a substantial room for improvement. As your security awareness program matures, the ability to make drastic changes reduces, and with this, you will need to adjust your strategic goals accordingly. The key to this process is to set goals that push your staff in a positive direction by improving the status quo without setting up predestined failure due to an unattainable goal.

How will we measure success? We need to identify and develop KPIs that will accurately measure the progress of our security awareness program. Based on these KPIs, we should be able to gauge the success of our marketing efforts in promoting our security awareness program and its underlying cybersecurity culture. We will delve deeper into KPIs later in the book as we build our security awareness program. An essential first step after creating our KPIs will be to establish a baseline using our current statistics to understand the current state of our security awareness program.

Set The Baseline

Strategic goals serve to establish the desired destination of your marketing activities, supporting your security awareness program. Knowing where you want to be is essential, but it is equally imperative to understand your baseline to comprehend where you are starting from when setting your strategic goals. If we do not know our current baseline, it will be difficult to develop realistic goals that can be achieved within the specified timeframe. A solid baseline is the foundation for defining our goals and milestones based on the current state of the security awareness program. Analyzing this baseline provides insight into areas to target, which can be utilized to formulate marketing plans that achieve the goals of the marketing strategy.

Reinforcing The Brand

Brand loyalty is a marketing term that describes consumers who repeatedly purchase the same product. This repetitive purchasing is driven by the brand's ability to consistently meet the consumer's requirements while eliciting emotions of a personal connection with the brand. When consumers are passionate about a brand, this can establish a robust customer base that provides a competitive advantage in the marketplace.

What is the primary goal we hope to achieve through our marketing efforts? The simple answer is to enhance our organization's security culture by increasing the ability of our employees to consistently exhibit the desired security behaviors. The underlying motivation driving this improvement in behavior is rooted in the emotional connections made through the marketing campaigns, which aim to boost brand recognition and awareness of the security program. Elevating our brand awareness and promoting the habitual adoption of our security best practices fosters a stronger cybersecurity culture, which ultimately reflects the strength of our brand loyalty.

We previously touched on building a brand for your cybersecurity program. This is important, and reinforcing brand recognition should be a continual part of your strategic goals. Several other strategic goals are relevant to all security awareness programs. These strategic goals are closely related and intertwined, so they often benefit and complement each other's marketing activities.

Increasing Market Share

We previously defined market share and its relationship to the degree of control a product holds in a specific market. A common goal of marketing strategies and associated campaigns is to increase a product's market share by taking customers away from the competition. We are selling our security awareness program to employees within our organization by developing a marketing strategy to capture market share from our competitors. So, in this case, who is the competition we are going to take market share away from?

This is where I take an innovative approach to defining the market for our security awareness program and its underlying cybersecurity culture. I view an organization's cybersecurity culture as a marketplace controlled by two competitors, each with its own market share. My security awareness program maintains the first market share that includes individuals within the organization who consistently demonstrate the skills and behaviors we expect from those who have adopted our cybersecurity culture. Conversely, the competing market share, which we are trying to capture, consists of employees who do not routinely exhibit the desired behaviors and, thus, have not fully adopted our cyber-security principles.

This competing market share, which we are attempting to garner "customers" from, includes individuals who routinely fail simulated phishing tests, do not complete training on time, or do not apply the desired security practices of our cybersecurity culture. As a marketer, your challenge is to develop marketing campaigns to reach those employees and to convey the personal benefits of your security awareness program to persuade them to become "customers" (part of your security culture). Marketing is about creating motivation within an individual that drives them to purchase our product, or in our case, demonstrate the desired secure behavior. Through our marketing efforts, we must tap into the motivating factors that encourage our employees to consistently exhibit the desired behaviors of our security culture. By consistently increasing the number of employees who perform the desired behavior, we will increase our market share and, in turn, improve the organization's cybersecurity culture.

How do we measure our market share? Market share is defined as the percentage of the marketplace that a product or brand controls. How do we identify the market share of our security awareness program? Later in this book, we will examine a unique metric to measure your engagement rate, which provides a method to determine the current market share of your security awareness program.

Changing Employee Behavior

Employee involvement is a critical component of all organizations' cybersecurity defenses.

An essential element of the marketing strategy is to improve the desired behavior of employees, which has a direct impact on increasing our market share. This behavioral improvement increases the security and protection of the organization's resources by reducing risky behavior. As employees adopt and demonstrate the desired behaviors of the cybersecurity program, they become an active part of the organization's cybersecurity defenses. The improved employee behavior reduces risk and additionally contributes to the marketing strategy's proposition value by improving the organization's cybersecurity culture. As employees adopt and habituate the desired security behavior, they become integrated into the organization's cybersecurity culture.

Measuring changes in employee behavior may be accomplished through different metrics, but three commonly used metrics are click rates, reporting rates, and training completion rates. When we reach "Measuring Success" later in this book, I will provide you with some unique ways to measure these statistics in your security awareness program, which can offer valuable insights into the performance of your marketing efforts.

Improve Cybersecurity Culture

Developing a strong cybersecurity culture is the fundamental goal of all security awareness programs. Improving employee behavior to align with the organization's desired behavior is essential for establishing a mature cybersecurity culture. Continually demonstrating the desired security practices fosters a cybersecurity culture, as individuals typically emulate the actions of their peers. This resembles word-of-mouth advertising, where the virtues of a product are spread by those who have experienced tangible benefits from using it. We want our employees to believe in our product, and in doing so, we create an environment where employees expound the virtues of our security awareness program to their peers. This peer recommendation to perform the desired security actions ingrains this behavior into the organization's fabric. Over time, this peer reinforcement helps establish and solidify the organizational norms that comprise the security culture.

Take it with a grain of salt

Measuring culture can be difficult. You may choose to incorporate culture surveys into your cybersecurity program. Surveys can be effective, but I have discovered a couple of caveats when conducting them. Employee participation is crucial to creating an accurate representative sample that can be used to develop credible results. If you choose to conduct the survey voluntarily, you may only receive feedback from a portion of your workforce, which may not accurately represent the full population. A common problem under this scenario is that you only receive feedback from employees who have either extremely positive opinions or extremely negative opinions. Employees who are indifferent in their view of the security awareness program may not provide any feedback, which leaves a significant void in the survey's results.

Employees may not trust the anonymity of the survey. If employees believe their answers are not anonymous, they may provide less than truthful responses due to fear of possible retribution. Another issue is that employees use the survey to provide feedback on more than just the cybersecurity culture. Depending on how closely Security Operations are associated with IT Operations, respondents may provide answers that encompass the entire IT department, rather than being solely focused on the cybersecurity culture. We have observed this type of result while reviewing the responses of users in a free-text comment field provided at the end of the cultural survey. Based on the respondent's feedback in comments, we could ascertain that the survey may have been completed with some bias due to their interactions with other areas of the IT department. These issues skew your results and produce an inaccurate depiction of your cybersecurity culture, nullifying the usefulness of the feedback in accurately measuring the organization's cybersecurity culture.

These caveats do not preclude the use of security culture surveys, as they are a valuable tool that provides insight into the current perception of an organization's security culture. However, it is essential to evaluate the validity, completeness, and accuracy of the results to understand their actual value.

Building Marketing Plans

The marketing strategy serves as the framework that provides guidance and objectives for building the marketing plans. The marketing plans serve as stepping stones, outlining the intention behind specific marketing efforts in support of achieving the marketing strategy's goal. The marketing plans are where you get to develop the marketing campaigns that are going to persuade your employees to become active participants in your cybersecurity program and culture. Our marketing efforts, to some extent, will always promote the adoption of the organization's desired secure practices.

Establishing a Connection

Effective communication regarding the benefits of your security awareness program begins with establishing a connection with your employees. The marketing plans and associated marketing campaigns provide a platform to creatively engage and capture the attention of your employees, highlighting the benefits of participating in your security awareness program. A key to making strong connections with your employees through your marketing campaigns is to develop messages that resonate with them. Craft your communications to elicit an emotional response that motivates participants to take action and engage in your cybersecurity program. The ultimate goals of your marketing efforts are to establish trust with the Security Operations team, promote the desired secure behaviors, and improve the organization's cybersecurity culture.

Campaigns

Developing marketing campaigns to promote your cyber-security awareness program is challenging because there is no tangible product to promote. This is where you can be creative by finding unique and interesting ways to engage with your employees and promote the benefits of the cyber-security program. Marketing campaigns are crafted around ideas, imagery, slogans, and messages that elicit positive feelings about your product. Your marketing campaigns are trying to educate and persuade your employees to adopt the organization's desired cybersecurity practices. Cultivate your campaigns in a fashion that details the personal benefits of these practices while advancing the goals of the marketing strategy.

Within the marketing plan, marketing campaigns are typically short-term events that often coincide with various holidays or special events. The duration of your marketing campaigns varies depending on the purpose of each campaign. Campaigns may be very short, lasting a single day, supporting events like World Password Day. More elaborate campaigns may span several weeks or months and involve various activities to promote events, such as Cybersecurity Awareness Month. Appendix A lists some key dates for events related to cybersecurity. Use these dates or identify other special days or holidays that align well with your organization's security awareness efforts. There may be opportunities to combine actual holidays with cybersecurity events, such as Halloween, which occurs during Cyber-security Awareness Month, or World Password Day, which occasionally falls on Cinco De Mayo. This is where you get to be creative, so use your imagination to find innovative ways to design campaigns that both entertain your employees and promote the desired behaviors that support your cyber-security culture.

While the marketing campaigns provide a creative element to promote your security awareness program, it is vital to understand your organization's identity and the makeup of your current market share. If you fail to under-stand the current makeup of your market share, you may risk losing this following through the development of campaigns that are not viewed positively. The recent Bud Light marketing fiasco illustrates how a marketing campaign can go drastically wrong and cause significant damage to the core market share. If your organization tends to be more conservative, you will want to avoid creating racy messages to promote your security awareness program. Instead, work with your marketing department to tailor your message to fit your organization's persona. All employees in your organization help shape the security culture, so avoid creating marketing campaigns that may alienate a portion of your workforce.

Know Your Limits

Push the boundaries and express your creative side to draw engagement to your security awareness program through your marketing campaigns. The marketing campaigns and associated materials provide you with a canvas to explore and develop creative ideas to promote your security awareness program, but this creativity must be tempered with an understanding of the core market share and the desired perception of the organization. As you begin developing campaign ideas, it is essential to thoroughly understand the values of your core market share to ensure your message aligns with these values. In

---

our efforts to gain new market share, we do not want to risk losing our existing market share by introducing campaigns that may be perceived as offensive. The campaigns additionally must account for the persona and public image the organization wants to portray.

Examples:

Several years ago, Palo Alto Networks ran a campaign to promote its CNAPP product using a tagline of "Shift Happens." This campaign focused on adopting a shift-left mentality to encourage the introduction of cybersecurity earlier in the application development process by using Palo Alto's CNAPP product. The tagline was a fun play on the mainstream phrase "shi* happens".

Splunk is another technology company that promotes its products through creative slogans on T-shirts. These slogans, such as "Take the SH out of IT" or "You bet your sweet SaaS," are fun word plays.

The examples shown above are acceptable for technology companies that may be a little more risqué with their marketing, but may be considered offensive by some individuals. Know your organization and your target audience to determine which types of marketing campaigns and messages are acceptable. Understanding these thresholds enables you to develop creative marketing campaigns, reinforce your brand identity, and capture new market share while maintaining your current market share and the desired corporate image.

Reference

https://merchandise.cisco.com/splunk.html

Summary

The purpose of the security awareness program is to increase the organization's security by improving the employees' security behavior. Market the benefits of the program to educate employees about cybersecurity threats and their role in protecting the organization's digital assets. Develop marketing initiatives for the program, including long-term and short-term plans with achievable goals. Promote the identity of the security awareness program to build brand recognition among your workforce. Increase the program's market share by changing employees' behavior to strengthen the cybersecurity culture.

Additional References

Brand Loyalty
What is Brand Loyalty - Definition, Importance with Examples | Marketing Tutor
https://www.marketingtutor.net/what-is-brand-loyalty/

Marketing Plan
How to create a marketing plan in 2025
https://www.smartinsights.com/marketing-planning/create-a-marketing-plan/how-to-create-a-marketing-plan/

Marketing Plan: Types and How to Write One
https://www.investopedia.com/terms/m/marketing-plan.asp

Marketing Plan
https://corporatefinanceinstitute.com/resources/management/marketing-plan/

Branding
Branding
https://www.ama.org/topics/brand-and-branding/

What is Branding: Branding Explained - Everything You Need to Know to Build a Powerful Brand
https://www.brandvm.com/post/what-is-branding-branding-explained

5

Developing Your Security Awareness Program

Develop your security awareness program from an employee-centric perspective and explain why it is beneficial for employees. Marketing is about reinforcing the benefits of a product or service, so continuously remind your employees that the security awareness training provided not only protects the organization but also equips them with valuable tools and techniques to protect themselves in their personal lives. You are marketing an intangible lifestyle product aimed at enhancing your organization's cybersecurity culture. If your workforce adopts the promoted security principles into their lifestyle, which translates into behavior changes, they will become more secure at home and at work.

What's In It For Your Employees?

A good starting point for understanding how to market your security program to your workforce is to ask, "What is in it for them?" As you begin to answer this question, you reveal ideas and strategies to promote your security program that resonate with your colleagues. This provides the impetus for them to "purchase your product," which is active participation in your security awareness program, which leads to a strong cybersecurity culture. How you market your security awareness program to your employees can significantly impact their participation and engagement.

Market the benefits of participating in the security awareness program at the personal level in addition to the organizational level. Employees are required to complete numerous training courses throughout the year, with no explanation of potential personal benefits, so the training is viewed as just another task to complete. Change the employee's perspective on your security awareness program by explaining how their participation will not only help the organization but also protect them in their personal lives.

Adopting a personal viewpoint allows you to promote cyber-security principles in a way that encourages individuals to not only learn how to protect the organization but also themselves from malicious actors.

When I begin to explain the benefits of incorporating the desired security practices into an individual's lifestyle, I like to point out that, at home, their only defenses are likely a basic firewall provided by their internet service provider and perhaps an antivirus application. There will be some individuals who are tech-savvy and have stronger defenses, but I have found that the majority typically do not have a solid set of defenses on their personal devices. This immediately gains their attention as you are now offering additional security measures to protect their personal information. The promotion and encouragement of sound security principles in all aspects of an individual's life initiates the transformation and adoption of the desired behavior, thereby strengthening their security culture.

Personalize Your Message

How does cybersecurity affect the individual? What is the impetus for them to actively participate in the organization's cybersecurity program? If there is no personal gain from involvement in your security program, employees may lack motivation to perform the desired behaviors. You must craft your messaging to explain how adopting the requested security practices directly affects the individual's well-being.

Most individuals do not connect all the dots to understand how a data breach against an organization will personally affect them. It is your responsibility to provide this information in a manner that captures their attention and raises awareness of how their active participation in the security awareness program benefits them personally. In addition to providing methods to protect themselves from potential personal data breaches, we continually reinforce with our workforce that their participation in our security program positively impacts their livelihood and the livelihood of their colleagues. We explain this by detailing the average cost of a data breach. This explanation continues by stating that although we have cybersecurity insurance, it will only cover part of the overall costs. Non-critical operating expenses, such as bonuses, wage increases, and extracurricular activities, including employee lunches and outings, will cover the remaining costs. This highlights how a data breach against the organization can affect them personally.

Examine data breaches against organizations in your industry or of a similar size. Detail to your employees the trials and tribulations the organization's staff went through as they recovered from a data breach. This might include explaining how processes typically handled through computer technology were being performed by hand. Many employees take for granted the computer technology used to perform their daily tasks until the idea and thought of conducting those processes by hand is presented.

The idea behind this is not meant to be a scare tactic to get employees on board with your program. The reality is that most employees do not consider the potential ramifications of a data breach on the organization and how it might directly affect them. Find ways to facilitate the desired behavior by personalizing the benefits and merits of supporting the organization's cybersecurity program.

Avoid Becoming A Victim

Encourage your employees to actively participate in your security awareness program, as it can help them avoid becoming a victim of a malicious actor. This training will not only help them prevent threats that put the organization at risk, but it can also equip them with the skills necessary to identify attacks that may affect them personally. As demonstrated time and again, the average individual often does not understand all the various methods employed by malicious actors to manipulate their behavior and facilitate a cyber-attack. Malicious actors are always searching for prey, so we must teach our employees about the tactics used to initiate an attack. Understanding the tactics and being able to identify them are essential first steps in avoiding becoming a victim. Attacks often begin by manipulating an individual's actions to coerce them into engaging with malicious content or exposing sensitive information.

Humans possess certain traits and qualities that are ingrained in our behavior, which highly skilled malicious actors know how to manipulate to initiate attacks. Manipulating a victim may be through face-to-face encounters, but occurs more frequently through electronic communications. Email represents the most significant initial attack vector for an organization, which may be used to distribute malware or expose user credentials. Additionally, employees are frequently targeted through text messages and phone calls. The malicious electronic communications attempt to manipulate your employee's behavior by invoking a strong call to action, a sense of urgency, or a negative consequence such as a legal threat, financial loss, or missed package delivery. It is imperative that your security awareness program adequately trains your employees to recognize traits and characteristics in communications that may indicate malicious intent.

Build It and They Will Stay

The average person is often unaware of the behavioral manipulation being conducted against them. The principles used to manipulate a person's behavior are not only utilized by the bad guys, but you can often find examples of manipulation in innocuous applications.

I admit that I have a slight addiction to the game SimCity BuildIt and spend more time than I should creating the most monumental cities. This game boasts a diverse range of characteristics and features that engage players, making it highly addictive.

The game incorporates multiple timers with varying durations to complete tasks. The activities associated with these timers reward you with Simoleons or items for your city if completed within the specified time frame. Additional pop-up events appear with their own timers, persuading players to stay in the game and complete the associated task. The timers create a self-imposed deadline: if you want to receive specific items, you must play the game to reach those goals before the timer expires. Essentially, the game's creators attempt to manipulate the player's behavior to keep them engaged by leveraging human nature and motivation to elicit a reward.

Figure 5.1: *SimCity BuildIt*
Source: Vera Aksionava/Shutterstock.com

The use of timers in SimCity BuildIt creates a sense of urgency to complete the task. Creating a sense of urgency is a common tactic used by malicious actors to persuade victims to take action to avoid a negative conse-quence. In SimCity BuildIt, you receive a unique item that can be incorporated into your city, but the underlying motivation remains the same. Educating employees on how to identify techniques that create a sense of urgency can significantly help them defend themselves against malicious actors.

By no means am I saying the creators of SimCity BuildIt are doing anything wrong; they are simply using human behavior to keep players interested in the game. The point of this anecdote is to illustrate how easily and unnoticeably manipulation of human behavior can occur if you are unaware of the techniques. Luckily, in this instance, the player only loses several hours of their day and not their life savings.

Malicious actors utilize similar manipulation techniques to pull victims in through illicit responses to emails, text messages, and phone calls. Prepare your workforce for this type of threat, not only in the workplace but also in their personal lives. If your employees are ill-prepared to identify and avoid these tactics, the outcome may be game over.

Most individuals have likely received a fake text message claiming a fraudulent credit card transaction or a package that is out for delivery. We live in a society where everyone likes to act quickly, so when a notification of this nature arrives, people often act without verifying the details.

This swift, emotion-based response is what the bad actors are counting on to perpetrate their scam. Educate your employees on how to slow down and remove emotion from their responses. This simple action will help prevent them from becoming another victim.

Take Your Work Home

We often hear from our employees about loved ones who have been preyed upon by online predators, and the tactics they used in the ruse. This is particularly common among employees with elderly parents, who tend to be more susceptible to scams, but it also applies to younger family members. The Federal Bureau of Investigation's 2024 Internet Crime Report states that 333,981 complaints of internet fraud were reported, resulting in losses of $13.7 billion (US dollars). This report also states that individuals over 60 years of age filed 147,127 complaints, resulting in losses of $4.8 billion (US dollars).

Many times, employees inform us that they recognized the signs of a scam after receiving the training we provided. These unfortunate situations reinforce the need to personalize the message of our security awareness program to increase adoption. By adopting the security practices into their personal habits, our employees are empowered to identify and avoid potential online threats. Encouraging your employees to teach their family members the principles and techniques they have learned from your program instills a commitment to the desired behavior. This also follows the adage that "the best way to learn is to teach." Teaching the principles and techniques of the security awareness program fosters a more profound understanding of the prescribed concepts, leading to active participation in the cybersecurity culture.

References

FBI Releases Annual Internet Crime Report
https://www.fbi.gov/news/press-releases/fbi-releases-annual-internet-crime-report

Building Blocks of the Program

All security awareness training programs will have common elements, such as simulated phishing campaigns and training. How you choose to implement these elements can have a profound impact on whether your program is designed to fulfill an audit checkbox or distinctly improve the overall security of your organization.

The purpose of the security awareness program is to promote and develop a strong cybersecurity culture within the organization. Employees can be one of the most effective elements of your security stack with proper training and motivation. Build your program to challenge your employees to hone their skills as part of the organization's cybersecurity defenses. The program is essentially a patch management system for the human element of your security defenses. This is a long-term proposition, as the security awareness program must continually enhance all employees' skill levels while being agile enough to adapt to the ever-changing threat landscape.

Measuring and tracking changes in the adoption of the program's security practices and the organization's cyber-security culture is a challenging task. The building blocks of your security awareness program, if designed correctly, can provide a robust set of metrics that offer valuable insights into the overall state of the security culture. Reviewing these metrics regularly allows you to identify weaknesses and adjust the program to mitigate them.

User Groups

The first building block is user groups, which serve as the cornerstone of your security awareness program. Developing user groups is a crucial first step in a successful security awareness program, as this element will be used to establish your simulated phishing campaigns and provide a means to create metrics to measure progress in your cybersecurity culture.

Remember, we are looking at our security awareness program through a marketing lens, so the user groups effectively represent the target demographic for our marketing campaigns. This consideration may influence how you create your user groups. Factors to consider are departments or job functions that may require specialized or alternative training as part of their program. If your security awareness program serves a multinational organization or there are significant regional differences within the workforce, these factors may be considerations when developing your program's user groups.

Tenure Groups

A foundational element of my security awareness programs revolves around a special set of user groups I term *Tenure*

Groups. Tenure Groups are defined by the length of time an individual has been part of the organization. I found success with defining my groups based on **Table 5.1**. Based on your requirements, you may want to adjust the timeframe or the number of tenure groups.

Table 5.1: Tenure groups

Group Name	Tenure
Year 0	Less than one (1) year of employment
Year 1	One (1) to two (2) years of employment
Year 2	Two (2) plus years of employment

The Tenure Group is a critical component in the design of my security awareness training programs, as it identifies which simulated phishing campaigns users participate in. This information provides metrics that measure the effectiveness of our security awareness training programs and the adoption of our cybersecurity culture.

Specialty Groups

Your security awareness program may incorporate "Specialty Groups" that are used in addition to the Tenure Groups. Specialty Groups may be based on factors such as work location or department.

Examples of potential Specialty Groups are:

- **Application Development** – this may involve providing instruction on secure coding and organization practices involving application deployment.
- **Human Resources** – which may require targeted training to understand techniques to conduct and spot potential fraud when interviewing a job candidate remotely.
- **Outside Sales** – by the very nature of this position, they will likely not come into the office regularly. This will present a challenge for developing the corporate security culture in this group. Additionally, they are typically working from remote areas such as home, coffee shops, and hotels, which require training to work safely in those environments.
- **Work From Home** – this phenomenon grew dramatically during the COVID pandemic and is now part of standard business practices. It is not uncommon to find work-from-home users working from areas other than their homes, including hotels, relatives', and coffee shops. This necessitates training these users about potential threats found in these locations and how to work safely in remote environments.

Phishing Campaigns

Train for the fire you hope you never need to fight. I say this because my office is located next to a fire station, and I routinely see the firefighters practicing and training their skills. In watching this, I began to equate their training to how I develop my simulated phishing campaigns. The repetition in the training firefighters undergo creates muscle memory, so when the time comes, they instinctively know how to handle the situation without thinking or letting emotion get in the way. This enables a methodical and pragmatic approach to handling a potentially stressful situation.

Your employees need to be prepared in the same manner to identify and handle threats that reach their inboxes. The training provided by your simulated phishing campaign should routinely expose employees to real-world threats, so they are skilled and prepared to handle the threat calmly, as prescribed by the security operations team. Enhancing your employees' skills to identify and handle potentially malicious content reduces the risk posed by this type of threat. This risk reduction reinforces their role within the organization's cybersecurity stack.

Simulated phishing campaigns should be an integral part of every security awareness program, but their effectiveness in preparing employees to handle potentially malicious emails depends on how they are implemented. These campaigns serve several purposes, acting as a continual form of training to prepare employees to face real-world threats and as a consistent evaluation of an employee's performance and adoption of the security program's practices and principles. Analyzing the employee's performance allows you to adjust the program based on observed weaknesses or risks. In developing effective simulated phishing campaigns, I have found several key components that make training more effective and measuring weaknesses in your program easier to identify.

Content Matters

The content of the simulated emails must reflect the types of emails an individual could expect to receive during their everyday activities. Simulated emails that blend in with regular emails provide a much more accurate measurement of the employee's ability to identify and report suspect emails. If the emails contain information that is easily identifiable as a test, or if the information in the email is irrelevant to the individual's job, then this will skew the test results. If the simulated phishing campaign results are not accurate, then they are not a reliable measurement of your employees' adoption of your desired security practices and overall security culture.

Creating realistic email content can be a challenge, especially when an organization has a culturally diverse workforce or highly specialized departments. Most platforms that host email simulations provide templates that can be tailored to present email content that precisely reflects the culture of the targeted group. Using these templates can provide more accurate results because the simulated emails will adjust their content to blend in with the typical content of the employee's mailbox, taking into account any cultural differences. If your organization has highly specialized departments, it may be necessary to create simulated templates specifically tailored to the content of those departments.

AI is a game-changer for creating highly targeted simulated phishing email content. Many simulated platforms now offer the ability to create content that utilizes AI to target your end users. Malicious actors are already using these tools to craft emails targeting your employees, so the days of identifying phishing emails by bad grammar or poor content are gone. Adapting this technology into your training campaigns can only reinforce your employees' ability to identify and report suspect email content effectively. Incorporating highly targeted, AI-driven content into your simulations helps prepare and train your employees for highly sophisticated attacks, as the prolific use of AI by malicious actors is expected to increase over time.

One use of AI we have noticed an increase in activity around is what we term non-technical phishing emails. We define non-technical emails as phishing emails that contain no malicious content in the links or attachments. The lack of any malicious content circumvents many of our email security controls that attempt to identify compromised payloads. The malicious content is presented as the actions the body of the email prompts the end user to take. This is often an invoice scam in which every element of the email, including the attachments, appears legitimate, except that the payment information is a bogus account. These types of scams are particularly tricky because they rely on the recipient to verify the content's legitimacy before paying the attached invoice. New AI tools are available to counter this type of attack, but instances will still occur where malicious emails slip through. This is where a properly trained staff can make the difference between preventing an attack and becoming another victim of fraud.

The efficacy of our metrics is directly driven by providing more sophisticated and realistic content in your simulated email campaigns. If we cannot accurately measure the program's performance due to skewed metrics, then we will be unable to correctly identify weaknesses and make necessary adjustments to mitigate our findings. The importance of realism in our simulated phishing campaigns cannot be understated. If our program does provide realistic simulations, the entire effort will be a waste of resources.

Difficulty

Setting the difficulty of your phishing campaigns is a key component to consider when designing your program. This is one of the areas where the "Tenure Groups" discussed earlier come into our program. I recommend setting up three distinct difficulty levels (moderate, complex, and everything goes). These difficulty levels are then applied to the Tenure Groups (see **Table 5.2**). The idea behind the progressive difficulty is to build the employee's confidence by providing initial success, then reinforcing this continued success by increasing the difficulty as the employee's skills improve. A key consideration is that if employees experience immediate failure, they may withdraw from your security program, feeling embarrassed or discouraged by their initial results. This is a training program, so you want to build the skill set within your employees. To do this, you need to establish a solid understanding and foundation to build upon.

Table 5.2: Phishing campaign difficulty

Group	Difficulty
Year 0	Moderate
Year 1	Complex
Year 2	Everything goes

Matching the simulated campaign's difficulty level to employees' tenure allows the program to escalate the challenge in line with the skills the employee develops throughout the program. This begins with the Year 0 group. This group may include employees in their first professional job, so they may not have prior experience with this type of security training. The Year 0 group may consist of employees who have had prior security training, but that training may lack the thoroughness or depth of your program, so their skills may be insufficient. We want our simulated campaigns to introduce this group to realistic challenges that build their foundational skills and ability to identify the characteristics of a malicious email. The simulated emails sent to Year 0 are designed to challenge their ability to recognize malicious breadcrumbs in emails, while also allowing them to achieve some early success to boost their confidence. We aim to introduce them to the organization's security practices and principles, so they become an integral part of the security culture.

As employees progress through your simulated training program, the emails become more challenging to increase the employees' skill level. When an employee progresses to Year 1, we increase the difficulty of the simulated phishing emails, making the breadcrumbs more challenging to identify. We are still controlling the content to reinforce the principles further and build upon the employees' experience. Upon reaching the Year 2 group, we believe employees should have the expertise and knowledge to identify and handle the most challenging simulations we can generate. This allows the security team to remove all restrictions on the training and present campaigns as realistic and diabolical as those of real malicious actors. This buildup and continual exposure to the ever-changing threat landscape prepare the employees to be a strong component of the organization's overall cyber-security defenses.

Lombardi Moment - Practice Makes Perfect

I was once asked why I believe changing the difficulty in our simulated phishing emails should be part of the security program.

I often use a sports analogy when discussing the theory behind gradually increasing the difficulty of simulated phishing emails as users progress through the program. I have played a variety of sports throughout my life and have found that I become a better player by facing stronger competition. Professional athletes did not suddenly arrive at the professional level; they spent many years practicing, training, and competing against increasingly more vigorous opponents. The same principle applies to the progression of difficulty within the simulated phishing campaigns. Your simulated phishing program should continually challenge your employees over time with more difficult tests to increase their skills and ability to identify potential threats similar to those posed by real threat actors. As employees hone their ability to identify and mitigate these advanced threats, they become an increasingly effective component of your organization's security stack by reducing the human risk.

Randomness

Figure 5.2: *Randomness is an effective component of simulated phishing campaigns*

Source: Anthony Berenyi/Shutterstock.com

Randomness is a crucial element in the development of your simulated phishing campaigns. The simulated campaigns incorporate two distinctly different aspects of randomness. The first element we will explore determines if employees receive the same or different simulated phishing emails. The simulated campaign should send random emails to each user with varying difficulty based on the training group to which they belong. A potential caveat to sending the same email to all employees is introducing a cubicle prairie dog that warns those around them about what to look for in their inboxes. This alert will notify users who have not yet received the simulated phishing email, potentially affecting the campaign's metrics. Additionally, this precludes the need for those recipients to evaluate the email, thus an opportunity to improve their skills is missed. A second element of random-ness examines the delivery schedule for the simulated phishing campaigns.

Campaign Schedule

The campaign schedule is a critical factor in the development of a simulated phishing awareness program. Phishing campaign schedules vary by organization and may be weekly, monthly, or yearly. If you are only sending simulated phishing campaigns monthly or yearly, this may not be suffi-cient to adequately prepare your employees to identify and detect real-world threats. I design my programs around a weekly campaign schedule, which is optimal for ensuring the participants in the program have ample exposure to different types of threats likely to appear in their inboxes. A higher frequency of simulated campaigns prepares your employees to be strong defenders through repetition and continual practice. The adage, "practice makes perfect," should be reflected in your campaign schedule, because the more simulated threats you present to your employees, the more adept they will become at identifying email-borne threats.

The increased frequency allows your program to adjust and adapt campaigns to evolving threats presented through email. Due to the frequency and evolution of threats, limiting your schedule to monthly or yearly makes it impossible to show your employees the full range of threats they face. The evolution of threats has increased exponentially with the introduction of AI. Malicious actors are now using AI to develop new phishing threats with a renewed sense of realism and more targeted spear phishing qualities. Identifying malicious emails based on bad grammar is now a thing of the past. Exposing your employees to simulations that reflect the use of AI is imperative and should occur regularly to adequately train them on techniques to identify this type of threat.

Sending simulated phishing emails weekly is advantageous to achieving the desired results from your employees, but how do we define a week? Typically, a simulated campaign using a weekly schedule follows a typical workweek (ex., Monday – Friday). It is essential to define a workweek, as this will be used in comparison to the recommended campaign schedule. We will define a work week as Sunday through Saturday, allowing organizations to choose which days to exclude based on their regular operating hours. As part of your simulated phishing campaigns, you may decide not to send emails to employees on days your organization is not open for business.

The problem with using a traditional workweek schedule is that once an employee receives their training email for the week, they know they will not see another simulation until the following week. The knowledge that another simulation email will not be sent until next week presents the potential for the employee to lower their guard. Creating an unpredictable pattern of delivery with your simulated campaigns is an integral component of your security awareness training program. When designing simulated phishing campaigns, I prefer to use a schedule from Wednesday to Tuesday, which I define as the campaign week. This may seem like an unusual weekly schedule, but several key factors contribute to its effectiveness in keeping employees focused on reviewing their inboxes. A random daily delivery schedule should be implemented to prevent employees from becoming accustomed to receiving materials on the same day each week. Combining random delivery days with an offset campaign week creates a "chaotic" simulated delivery schedule that prevents employees from knowing if potential phishing emails they receive are real or training. This is achieved through the randomness of this methodology, whereby employees may receive one, two, or no simulated emails during a typical workweek (see **Figure 5.3**).

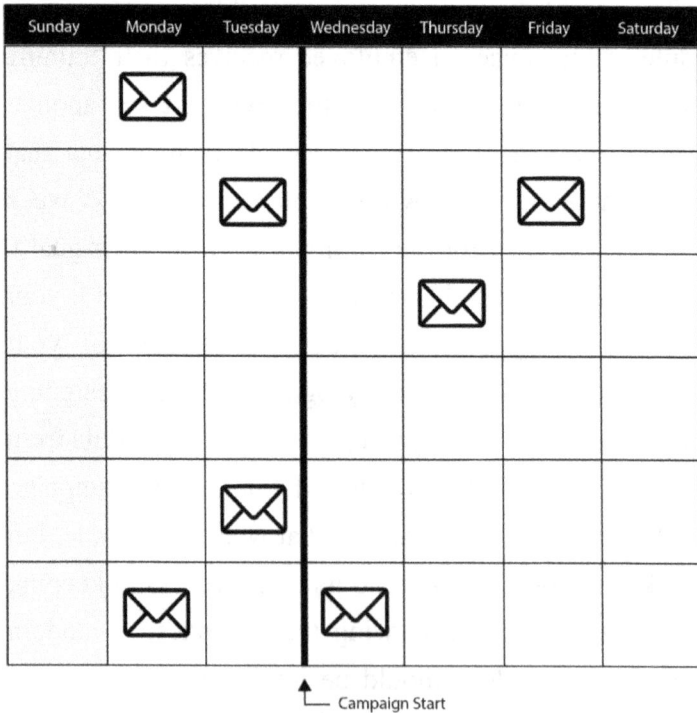

Figure 5.3: Campaign delivery example

Let's delve into this madness and explore how this works. The schedule described is based on a business being open Monday through Saturday. We know employees will receive one email between Wednesday and Tuesday. The delivery days are randomized, allowing delivery to occur on any day during the workweek. Using the delivery schedule depicted in **Figure 5.3**, let's review the delivery of simulated emails to understand how this offset delivery schedule works.

The employee receives a simulated phishing email on Monday from campaign A, which ends on Tuesday. Campaign B begins on Wednesday, and the employee receives this campaign's simulated phishing email on Tuesday of the following work week. Campaign C starts on Wednesday, and the employee receives this campaign's simulated phishing email on Friday. In this scenario, the employee receives two simulated phishing emails during the same workweek from two different campaigns. Campaign D then sends the next simulated phishing email on Thursday of the following work week. Campaign E begins on Wednesday, but does not send the employee's simulated phishing email until Tuesday of the following work week. In this situation, the employee went for one work week without receiving any simulated phishing emails. Campaign F begins, and the random delivery cycle continues. This erratic delivery schedule forces your employees to review all emails. This random pattern, combined with the potential for zero, one, or two simulated phishing emails during the workweek, prevents employees from letting their guard down. The need to continually evaluate all emails received helps drive the desired behavior in our employees to utilize their skills in identifying and reporting potential threats.

Gotcha Moments

"It's a trap" - General Ackbar.

No, it's not a trap; it is a teachable moment.

Remember, we want to build trust with our workforce, so we don't want them to feel like we are trying to trap them with the simulated phishing campaigns. Simulated phishing campaigns should not be treated as "gotcha moments." It is essential to remind your workforce that simulated phishing campaigns are training exercises designed to elevate their skills and enhance their security awareness. The goal is to prevent employees from interacting with simulated emails. Security Operations celebrates when a training campaign has zero clicks, not when a large number of employees interact with a test email. If an employee does make a mistake on a simulated phishing email, the good news is that it is a controlled simulation. Be positive and use this as an opportunity to teach the employee about the identifiers that should have alerted them to the nature of this email. We do not want them to make the same mistake when a real phishing email reaches their inbox.

Phishing Derby

The only exception to the "Gotcha Moment" would be if you are hosting an event such as a "phishing derby", and the employees are aware of the contest. Phishing derbies can be entertaining events that generate a lot of chatter about the cybersecurity program, as employees compete to win the competition. If you are unfamiliar with phishing derbies, it is a contest that your employees are fully aware of, and acknowledge that the security operations group will be sending more frequent and trickier simulated phishing emails. We have commonly conducted this type of contest in combination with our Cybersecurity Awareness Month promotions.

Figure 5.4: Phishing derby Ghostbusters landing page

One year, we used a Ghostbusters theme for Cybersecurity Awareness Month, and as part of our activities, we conducted a phishing derby with our staff. If an employee clicked one of the phishing derby emails, they were greeted with a pop-up message stating "You've Been Slimed" (see **Figure 5.4**). We had great employee feedback from this campaign. Employees routinely stopped us, so they could tell which emails we "slimed" them on. These conversations about being "slimed" were typically upbeat and fun because our employees knew it was a contest that was deliberately trying to trick them. The cheerful nature of these conversations was due to the trust we had gained with our employees and the acknowledgment that, under normal circumstances, we do not treat these failures as a "Gotcha Moment."

Phishing Fatigue Is a Myth

I often hear of organizations and security professionals that do not want to send "too many" simulated phishing emails to their employees due to "Phishing Fatigue." What constitutes "too many" emails? The average employee receives approximately one hundred twenty (120) emails per week. Does receiving one or two training emails during the week really overburden your employees? I do not endorse the practice of sending simulated phishing emails to employees on a daily basis, as that would be excessive. However, based on the current threat landscape, employees do need to be adequately prepared to defend their inboxes.

Employees must spend some time evaluating the simulated emails, which possibly takes a few minutes away from their production. If they have bought into your training and security practices, they are already evaluating every email that arrives in their inbox. Removing an email from their inbox requires little extra effort, such as reporting or deleting it. Most importantly, do you think the malicious actors are going to stop sending emails to your users because they have already sent five malicious emails this week and are worried about users having "Phishing Fatigue"? Considering malicious actors send an estimated 3.4 billion phishing emails daily, according to ZDNet, it does not seem burdensome for your security team to send one or two emails per week to ensure employees remain vigilant in their role of defending the organization from cyberattacks.

According to IBM's Cost of a Data Breach, phishing remains the most prevalent initial attack vector, and this does not appear to be changing anytime soon. AI technology enables malicious actors to create high-quality phishing emails at scale. KnowBe4's 2025 Phishing Threat Trends Report Volume 6 reports that 84% of malicious emails contained AI-generated text or payloads. The threat to organizations from email is accelerating, not declining; therefore, employees who feel they are too burdened or fatigued by the additional practice of identifying these types of threats present a significant risk.

References

56 Email Statistics You Must Learn: 2024 Data on User Behaviour & Best Practices
https://financesonline.com/email-statistics/

2025 Phishing Threat Trends Report, Vol. 6
https://www.knowbe4.com/resources/whitepapers/phishing-threat-trends-report-6

Three billion phishing emails are sent every day. But one change could make life much harder for scammers
https://www.zdnet.com/article/three-billion-phishing-emails-are-sent-every-day-but-one-change-could-make-life-much-harder-for-scammers/

50+ Phishing Attack Statistics for 2025
https://jumpcloud.com/blog/phishing-attack-statistics

Several key considerations contribute to creating an effective simulated phishing campaign. The simulated campaigns are never meant to be "gotcha moments" but rather should be interpreted as teaching moments. Adjust the difficulty of the simulated phishing campaigns to continually challenge your employees and help them improve their skills. Randomize the delivery schedule to maintain a heightened level of awareness to prevent employees from lowering their guard. The simulated campaigns are a training tool to help your employees adopt the desired secure behavior.

Training Campaigns

Developing effective training campaigns for your employees is a crucial component in fostering participation and cultivating a robust cybersecurity culture. Training campaigns are one of the primary means of communicating the organization's desired security practices and exerting a significant influence over employee participation in your program. A key to creating successful training campaigns is developing campaigns that generate strong employee interest and engagement with the materials presented. If you do not generate employee interest in your training campaigns, the material may be overlooked and viewed as a required task. It is essential to understand your organization's demographics and desired cybersecurity culture to present training materials that resonate and generate interest from your employees.

Can I Binge Watch The Security Training?

This question has been asked by multiple employees who were participating in our security awareness training program. The quote resonates with the power of creating an engaging training program and explaining how it benefits the employees.

How many times have you ever heard an employee ask if they can watch multiple training sessions because they enjoy them? Most training programs repeat the same content year after year, making slight adjustments for new regulations or requirements. Employees typically view these training sessions as a chore that must be completed to fulfill an organizational or regulatory requirement. The content tends to be monotonous and does not engage the employee personally, nor is there any explanation regarding the benefits of the training (no personal connection). Frequently, the viewer's goal is to speed through the training and answer enough questions correctly so that they can move on with their day.

Let's explore some of the key aspects of creating a security training program that elicits excitement and eagerness from the participants.

Not Another Training Video

Most organizations require annual sexual harassment training, and employees often view this training as just another task to be completed as part of their job. When was the last time you overheard a fellow employee discussing how they were excited to complete their assigned sexual harassment training? It could be due to how the material is presented and the lack of explanation as to how it benefits the employee.

Let's approach the presentation of sexual harassment training from a marketing standpoint to help generate employee engagement. Imagine if the human resource department marketed sexual harassment training to employees as:

> *A training course that will help create and foster a comfortable and productive environment for all employees. Additionally, this training helps reduce organizational costs by avoiding unnecessary lawsuits, which can potentially provide additional funds for bonuses, pay increases, and other employee-oriented activities.*

Simply adding context and changing the tone makes this a very positive statement that explains the benefits to the employee and generates interest in participating. It makes it difficult not to want to be involved in this program, as it identifies several benefits of participation that will directly affect the employee. Most employees desire a comfortable workplace with opportunities for regular pay increases.

Apply this same approach to your security awareness training program by making the program personal to your employees. Making your content personally relatable to employees presents a more inviting proposition that drives engagement. Continuously promote your training program and the rewards it offers to each employee. When employees feel a personal connection to the training content, they are often more engaged and invested in the security awareness program because it is viewed as personally beneficial rather than a requirement.

Broadcast Programming

When deciding on the content that will make up your training campaigns, it is important to consider that individuals have preferences in their entertainment. You may ask, What does entertainment have to do with creating training campaigns? Your training campaigns are a component of your overall security awareness training program, which is a primary driver in establishing your organization's cybersecurity culture. Selecting training campaigns that your colleagues find entertaining will help with their engagement in your program and increase their comprehension of the materials you are presenting.

When deciding what constitutes entertainment, I look to television programming to understand that it encompasses a variety of different program genres, ensuring I have content that represents a multitude of entertainment preferences. We have previously discussed the role demographics play in creating your security awareness program; however, it is also essential to understand that individuals within these groups have different entertainment preferences. It is vital to recognize that a one-size-fits-all training program cannot be created. You must adapt the program to provide entertainment that appeals to all members of your audience.

Figure 5.5: *Broadcast program selection*
Source: Prostock-studio/Shutterstock.com

Taking a cue from traditional television programming, it's essential to fill your schedule with a variety of content to ensure you reach all segments of your audience. Individuals are unique and differ in their entertainment preferences.

Building your training campaigns from a television programming perspective allows you to schedule content that features various entertainment styles, such as comedies (sitcoms), dramas, documentaries (news), and trivia (game shows). Utilizing a diverse entertainment portfolio in your training campaigns ensures that your programming includes something entertaining for all members of your audience. This diversity provides your employees with an understanding that perhaps the current training modules are not their favorite form of entertainment, but another "show" will be coming up that is their preferred entertainment genre. The idea behind this diverse programming model is to present training content that keeps your employees excited and engaged in the program.

On Demand

As described, I have found developing my training program using a television programming model to be very successful and well-received by my employees. It provides a controlled structure to ensure all employees have been exposed to the same training. We, however, live in a world of on-demand entertainment, and many of the younger generations do not remember what it was like to have limited entertainment choices. To meet the request for on-demand entertainment, I like to provide optional content that is not part of the standard training program. This content may address particular topics or those with a limited shelf life. As mentioned, this content is optional, so employees are not required to participate in or complete any of this additional training. We do find ways to reward those who do complete this additional training through leaderboard recognition, bonus points on their scorecard, and prizes.

Specials

You may need to develop targeted training campaigns for specific groups within your organization. If your organization has an application development group, you may want to establish a supplemental program for that group that focuses on secure programming. Groups that work remotely, such as sales or work-from-home users, may have additional training related to the dangers of public internet. Evaluate the needs of your organization and develop specialized or custom content for specific user groups as required.

Prime Time

We have identified that making our training campaigns entertaining is an essential factor; however, we must also consider the frequency and duration of our training. I often hear of companies that conduct cybersecurity training once a year. This is great for meeting a regulatory requirement, but is this enough to ensure your users are prepared to adequately handle the evolving threat landscape? Increasing the frequency of your training campaigns is ideal, but you need to be cognizant of employee time commitment to your program, so you do not impede their job functions. In developing training programs with a higher frequency, I have found a tradeoff in utilizing shorter videos to be palatable to our cybersecurity initiatives, management, and employees.

The frequency of your training campaigns is a delicate balancing act, as you want to ensure your employees are prepared for potential cybersecurity threats without taking them away from their primary responsibilities for too long. It is essential to gain organizational buy-in for your program across all levels of management (we discuss building alliances in Chapter 6 – Promoting Your Security Culture). The goal of our training program is to keep cybersecurity threats at the forefront of employees' minds while they conduct their daily activities, ensuring these activities are performed safely. It is hard to argue against the concept of enriching employees' cybersecurity knowledge to protect the organization, but if it interferes with the employees' ability to perform their required duties, you will lose the support of management within the organization. Retaining manage-ment's support requires that you continually reinforce the benefits of your training program in protecting the organization, while demonstrating respect for the employees' time commitment.

The program I have found to be the most effective is a four-week cycle with training activities that require between eight and twenty minutes of the employees' time. We keep the training material short and vary the entertainment genre to retain employee engagement. This short duration is also palatable to our management as it does not take too much production time away from the employee. The schedule allows for a monthly reminder and reinforcement of the organization's desired cybersecurity practices without being too burdensome on the employees' time.

Every organization is different, so you must find that balance between the length and frequency of your training campaigns. When determining the appropriate schedule for your training program, be aware that you are competing for the employee's time with the production demands of their department and other training requirements, such as those from human resources. The training campaign is scheduled in accordance with management's buy-in for your security awareness program and commitment to enhancing the organization's cybersecurity culture. Build your alliances to gain support for your program and remind them that everyone is part of the organization's security stack. As part of the security solution, these components must be continually updated, just like other components that protect the organization.

Looking Forward

A mature security awareness program must be flexible and continually improve to adapt to new threats and evolving training techniques. New tools are emerging on the market that will help prepare your employees for the diverse threat landscape they face across all their electronic devices. I have recently viewed demonstrations of tools that conduct SMS text message simulations, much like the simulated phishing campaigns we incorporated into our programs for years.

AI technology offers limitless possibilities in creating innovative training methods that prepare our workforce to counter malicious actors. One technology I have recently previewed is interactive AI phone calls to test staff. These tests could target how the helpdesk handles password resets to prevent credential exposure, or ensure that call center representatives do not expose personal information. Continue to look forward with your security awareness program and identify new ways to prepare your staff for the threats they face today and in the future.

Summary

Make the security training relatable to employees and explain how using the prescribed training benefits them not only at work but also in their personal lives. Establish groups to assign and measure your security awareness training campaigns. Approach the implementation of simulated phishing campaigns by establishing a foundation of good practices and building employees' skills over time through repetition. Use training materials that employees find entertaining to avoid making it just another task to complete.

6

Promoting Your Security Culture

"Without promotion, something terrible happens...nothing!"
- P.T. Barnum

The goal of the security awareness program is to grow and mature the cybersecurity culture to meet the desired vision. Do your employees know what the vision is for the cyber-security culture? Do they understand their role in reaching this vision? If you do not actively promote the vision and the employee's role in achieving it, your cybersecurity program is destined never to achieve the desired results.

The security culture lacks a voice of its own. You must become the voice actively promoting the cybersecurity program's merits. The driving force for changing your organization's security culture must start with promoting the desired behavior we expect from our employees. Promoting this change in behavior is not a solo endeavor and will require the involvement of different alliances and peer influences to be effective.

Influencers

Influencers from a marketing perspective have been promoting products for decades, beginning with celebrity endorsements. In recent years, due to the phenomenon of social media, a new group of non-celebrity social media influencers has gained significant popularity and followings. Social media influencers have significant marketing clout, especially with the younger generations, because of their ability to establish a personal connection with their followers. This personal connection is built by sharing authentic, relatable experiences with a product or service. The ability to create this personal connection gives social media influencers the same, or possibly more, credibility than their peers.

Peer Influencers

One of the most influential groups that can promote your security awareness program is your peers, who can spread the benefits of participating in the program through word of mouth. Word-of-mouth advertising occurs when individuals share the benefits and positive aspects of a product with others based on their own experience. Word-of-mouth advertising is one of the most potent forms of advertising, as individuals place a high level of trust and objectivity in recommendations from those close to them. This plays nicely into the development of our security culture, which also relies heavily on peer influence.

Word-of-mouth marketing varies from natural word-of-mouth advertising in that promotion of the product to our peers is "seeded" by individuals associated with the product. Seeding is conducted by actively highlighting the product or service's benefits in conversations and other areas of influence. A key to successfully using this type of marketing is to be genuine and honest. The person who is "seeding" the conversation must believe in the product and its benefits. This type of marketing is commonly seen today with social media influencers.

In today's marketing, peer influencers are used to promote a wide range of products and services by "influencing" the buyer's behavior to purchase the promoted product. We are marketing our security program, so it is advantageous to cultivate groups within your organization outside of Security Operations to help promote and advance security awareness initiatives. This type of influence is found by creating alliances with other departments and forming a champion or ambassador program to encourage the adoption of your organization's desired security behaviors.

Build Alliances

Just like the show Survivor, you must establish alliances along the way to achieve victory. What is victory in terms of a cybersecurity culture? I define a victorious cybersecurity culture as a well-defined program that cultivates and drives employee involvement by increasing understanding of their role in defending the organization against cyber threats. Creating this victorious environment is challenging and requires different motivations for each employee. Motivating employee participation cannot be achieved solely by the Security Operations group, which is why it is critical to build and maintain strong alliances to be victorious.

Alliances that share your goal of establishing a strong cybersecurity culture bring diverse voices that help reinforce the objectives of securing the organization and how employees' behaviors affect this defense. Curating your organization's cybersecurity culture is a perpetual endeavor due to continual flux in the workforce and the ever-evolving threat landscape. Solid alliances ensure the desired security practices and principles that define the organization's cybersecurity culture are ingrained in the workforce throughout this lifecycle. The consistent reinforcement and education provided by strategic partners within the organization help employees to assimilate the desired behavior of the organization's cybersecurity culture.

As we discuss flux in the organization's workforce, you must stay aware of changes within the organization's management, particularly those that are key to your alliance network. As management changes in these groups, you will need to take the time to educate and cultivate relationships with the new personnel to ensure continued support for your cybersecurity program. This is especially true if the new management is from outside the organization, as they will not understand your organization's cybersecurity culture or its role within your enterprise.

Establishing key alliances is a critical component in garnering participation in the cybersecurity program, as these alliances often have significant influence over the organization's employees. Alliances are instrumental in establishing a mature program, and the following are some of the departmental alliances I believe are vitally important in making the organization the Survivor.

C-Level

The C-Level group is the most important alliance to establish in creating a well-adapted cybersecurity program. Many departments are instrumental in developing the organization's cybersecurity culture, but receiving the support of the organization's C-Level is an absolute requirement. A supportive C-Level exerts significant influence over the other alliances by underscoring the importance of building a strong cybersecurity culture within the organization. If the C-Level does not support the cybersecurity program's initiatives, securing the necessary commitment from the other alliances can be challenging. Securing C-Level support is arduous, but I have a couple of approaches that may resonate with them.

The cybersecurity program is not a revenue generator, but losing large sums of money due to a data breach is not either. Promote the benefits of a strong cybersecurity culture and how it helps reduce risk and protect the organization's data and corporate secrets. The protection of this information may not only be critical to the organization's success, but it may also be a regulatory requirement. If the organization's success revolves around proprietary trade secrets, explain how a strong cybersecurity culture can help prevent those secrets from public exposure, and the financial effects public exposure could have on the organization. If customer trust is vital to the organization, such as in financial services, highlight the brand-reputation damage a data breach could inflict by eroding customer trust. Recovering from brand reputation damage is a costly and challenging endeavor from which the organization may never fully recover. The key component of this message is to secure the support of the C-Level by highlighting the benefits of a strong cybersecurity culture and how it can help prevent revenue loss from data breaches or other cyberattacks.

Another approach I take with building alliances with our C-Level is that of "no news is good news." I often tell our C-Level that a secondary function of Security Operations is to keep them out of the news for all the wrong reasons. Given the high risk of data breaches today, the message resonates with executives who want to avoid negative publicity stemming from a data breach that affects the organization.

This helps create the communication channel and understanding that avoiding this type of event must involve properly training the human element within our organization. This helps generate support to foster a strong cybersecurity culture within the organization and provides the necessary backing to implement the cybersecurity program.

Marketing

The foundational principle of this book is learning to market your security awareness training and security culture to your organization's employees, so it should come as no surprise that building a strong, effective alliance with the marketing department is crucial to achieving success. You need to view your security program as a product and consider how to promote it to your employees so they will want to adopt the program's principles. Most highly successful products have strong marketing to promote them and raise consumer awareness of their benefits. Working with your marketing team, who are experts in promotion, you should be able to create campaigns to launch new facets of your program and maintain consumer awareness of your product, keeping your employees engaged with the security program.

If your security group lacks creativity, then enlist resources from your marketing department to help create fun advertisements that will draw your employees to your program. Marketing typically has resources that assist with graphical and layout design. Creating marketing campaigns is one of the best ways to promote your program, and it requires coordinating materials across multiple forms of media. If you are unfamiliar with the requirements of the various media types, your marketing department can provide guidance on developing a cohesive message that promotes your security program and encourages employee participation.

Marketing is a key group to build an effective alliance with. If your team lacks creativity, marketing can help create the graphics and other materials needed to promote your program. Marketing is likely the group that purchases promotional items (SWAG) for events that your organization participates in. We will discuss how SWAG plays a role in developing your security culture later in this book. The marketing department is where your organization's promotion and advertising experts reside, so they can provide guidance on how best to promote your security awareness program to your employees.

Staff Development

New employees typically begin their employment with no knowledge or exposure to your organization's cybersecurity culture. Staff Development is one of the first groups new employees encounter when they start their tenure with your organization, so you must have a strong alliance with Staff Development to ensure they promote the desired security culture from day one. You want to begin shaping and molding new employees' adoption of the desired security practices early, so they become ingrained from the onset of employment. Staff Development is an ideal group to help introduce cybersecurity culture, as they can promote best security practices while training employees on the functions of their roles. The introduction of security practices through staff development notifies new employees that the organization takes cybersecurity seriously and outlines the expectations for them to adopt the organization's promoted security practices.

Human Resources

Fostering a strong alliance with Human Resources benefits the security program in several ways. Human Resources assists in promoting the program to employees through internal channels and often handles disciplinary matters. We will touch on both, but let's begin with the positive aspects of promoting our program to employees. In many organizations, Human Resources coordinates and distributes internal communications to employees. These internal communications, such as employee newsletters, intranets, or mass emails, are the ideal platform for presenting the messages you have created with the marketing department. Human Resources may similarly influence employees, like Staff Development, making this a highly desirable alliance to cultivate.

When possible, I always recommend using positive reinforcement with employees to gain their support and help them adopt the desired behavior outlined in the security program. Positive reinforcement through rewards and recognition is much better received, and employees will typically continue with the desired behavior in hopes of receiving more accolades. Unfortunately, this positive reinforcement does not work with all employees, so it will be necessary to work with Human Resources to develop a program of consequences for continued and excessive failures. This is not the desired action, but remember that a single click on a malicious email can bring down the whole organization, affecting everyone's livelihood. Engaging with Human Resources to address employee behavior should be your last option, but ensuring that Security Operations and Human Resources present a unified front will demonstrate the organization's commitment to adopting a strong cybersecurity culture.

Department Heads

Creating alliances with department heads throughout your organization is an essential element in fostering the promotion and acceptance of your cybersecurity culture. In almost every reference on building a strong cybersecurity culture, it is stated that you need the support of your C-Level. We have touched on the importance of receiving C-Level support for your program, but I believe it may be more important to garner the support of your senior and middle management. The C-Level defines the security culture at a high level, but your organization's department management promotes the security program to the employees daily. This daily involvement with the employees provides continual reinforcement of the importance of adhering to the organization's cybersecurity practices. Cultivating a strong relationship with all levels of management is instrumental in a successful cybersecurity program. If the managers embrace the mission and methods to protect the organization, they will be inclined to encourage their employees to adopt the desired security practices.

Ambassador Program

Building a security ambassador or champion program within your organization is a great way to create a group of influencers for your cybersecurity program. Often, when you hear the word ambassador, it is in the context of governmental functions relating to communication between two different countries. Applying this to your cybersecurity program is not much different; rather than communicating between two countries, the ambassadors act as liaisons between Security Operations and other departments within your organization.

Ambassadors are not part of your Security Operations team; instead, they are individuals from other departments who may be interested in cybersecurity or in developing the organization's cybersecurity culture. One of the most significant benefits of fostering an ambassador program is that it creates a communication conduit between Security Operations and the frontline personnel in other departments. Much like governmental ambassadors that may need to translate between languages and cultures, your security ambassadors can disseminate security messages to their peers in a manner that is better understood and received by the group they represent. The ambassadors facilitate messages, free of jargon and easily understood by their peers, providing an effective channel for Security Operations to convey information about new security products, features, or initiatives to employees.

The communication channel with your ambassadors is bi-directional so that they can provide feedback from their respective groups on security initiatives and the overall cybersecurity culture. This type of input is invaluable for shaping your cybersecurity program. While Security Operations attempts to establish rapport with all employees, employees may be more open or upfront in discussing topics with their peer representatives rather than reaching out directly to Security Operations. The information provided by the ambassador's peer groups to Security Operations provides insight into how the employees perceive the security program and the effectiveness of the message. This unfiltered feedback provides highly influential insights into how to modify and reshape the security awareness program to retain or increase employee participation.

There is no right or wrong way to establish your ambassador program. Develop the program in a fashion that provides the maximum benefit to your organization.

Depending on the size and structure of your organization, you may find it beneficial to have multiple ambassadors to represent large departments or departments with distinct subsets. The program's membership should reflect the organization's entire workforce, and I suggest attempting to expand the ambassador program to include representation from all departments. Involving all departments in the ambassador program ensures you have a communication channel to effectively relay security initiatives directly to the entire workforce and gather well-rounded feedback on your security awareness program.

SWAG

Who doesn't like free stuff?

Promotional items, also known as swag, are marketing tools that have been used for years to promote products. When incorporated into your security awareness program, swag can be an effective way to increase participation. SWAG is designed to draw attention to your brand, which aligns with our previous review on why creating a logo for your security awareness program is so important. Promotional items often display the brand's logo, which should serve as a constant visual reminder of the product or service it represents. When properly incorporated into your security awareness program, it serves as continual reinforcement for users to adhere to and put into practice the desired security processes that create the organization's cybersecurity culture.

Disclaimer – Nothing Is Free

I often give presentations to high school students, and a common theme is that nothing is free. As part of my presentation, I use the example of a buy-one-get-one-free pizza offer. I explain that the second pizza does not just materialize; there is an actual cost for this pizza. The marketing department is absorbing the cost of the second pizza as part of the promotion, or the pizza shop is hoping to increase sales volume at a lower margin to make a profit.

BUY GET 1 FREE

Figure 6.1: *Buy one, get one free*
Source: Creative.factory/Shutterstock.com

The pizza offer has set the precedent that nothing is truly free. This now leads to examining the "free" apps the students are using on their phones. Once again, I explain to them that nothing is ever truly free and that there is a cost to the "free" apps they use. I proceed to demonstrate that their data passing through these applications is a cost for using the "free" app. This data, collected by the app creators, is being used or sold to third parties for a variety of purposes, including potentially malicious ones.

This exercise is meant to scare students, but, more importantly, to enlighten them about potential threats they are unaware of. The idea of getting something for free is constantly enticing, and malicious actors are always looking for a hook to draw a victim in. Look for opportunities to educate your employees about tactics and possible threats they may face online.

The success of promotional items is commonly measured by the number of impressions over the product's lifetime. Impressions define the number of times the promotional item, more specifically the logo on the item, is viewed by the end user. Higher impressions equate to more opportunities for the promotional item to remind the end user of the products or services behind the brand. The lifetime of an item determines how long the product will remain useful (i.e., T-shirts can only be worn so many times before they look too worn). Promotional items have an associated cost, and this is where some of our security awareness program budget will be spent. It is essential to maximize our budget, so identify promotional items with a long lifespan and that are useful to your employees. Items that are used more frequently and for more extended periods offer more opportunities for your brand to remind employees of the principles of your security awareness program.

Promotional items have varying lifespans, so it is crucial to understand the length of the campaign the promotional product will support. Products that support your general security awareness program should have a longer lifetime than those that support a specific event, such as World Password Day. A key component of extending the longevity of your promotion is identifying products that your fellow employees desire or find valuable.

Make It Collectible

One way to generate demand for your swag is to limit production. Limiting production inherently means that not all your workforce will receive a specific piece of swag. If the swag is highly desirable, it can create perceived value in your employees' minds, driving a fad-like demand. Fads driven by the perceived value of collecting rare items are not new and have been seen in crazes around Beanie Babies (**Figure 6.2**), Pogs, Webkinz, and Littlest Pet Shop, among others. At the height of the Beanie Baby craze in the late 1990s, these little plush animals accounted for a significant portion of eBay sales, and it was not uncommon to hear stories of individuals paying thousands of dollars for a supposedly rare animal to complete their collection. I am not encouraging you to create a fad or craze around your swag, but I am suggesting you tap into this human behavior to generate a marketing buzz for the items used to promote your security awareness program. The limited availability of highly desirable swag serves as a strong motivator, helping people perform the desired actions your cybersecurity program requires.

Figure 6.2: *Beanie Babies collectible fad*
Source: The Image Party/Shutterstock.com

Every year, we create a unique theme for our security awareness campaign during Cybersecurity Awareness Month and hand out various swag items related to it. The most sought-after item during this promotion is often our company-branded wearable merchandise (e.g., T-shirts or pullovers) that we produce each year. There are a couple of factors that make this a desirable item within our organization. The first is that we usually have a fun design that resonates with employees and draws attention to the campaign. The second driving force is that we understand that our employees desire wearables with the company logo that can be worn on Fridays. This creates strong demand for our paraphernalia, as new employees may not have any company-branded wearables, and tenured employees may be wearing out their own. The third element of our program is that we print only a limited number of promotional items, so not every employee will receive one, making them coveted among our employees. This directly ties to the fad mentality that creates perceived value for the swag, making it highly desirable among our employees.

How does this demand for our collectible wearables increase our employees' desire to perform the requested security actions? Two of our desired security behaviors we want our employees to exhibit are not clicking on malicious emails and reporting suspect emails to Security Operations for evaluation and extraction of indicators. As part of our Cybersecurity Awareness Month campaign, we recognize on our intranet those employees who have not interacted with a simulated phishing email for the last 18 months and have reported 75% or more of those emails to Security Operations. Every employee in this group automatically receives one of our wearables, and we make the presentation of these promotional items a big deal by going to their department, typically with our marketing videographer, to present the item to the employees. It is not uncommon for the members of a department to cheer for their fellow employees as they receive their prizes. During one of our presentations, an employee who was not receiving a wearable told me, "He hasn't clicked on email in months because he wants whatever we offer next year." This statement, along with the recognized employees' actions, provides strong evidence that the limited, desirable aspect of the swag is driving participation in our security awareness program. On occasion, we had to increase the reporting rate to 80% because we did not have enough promotional items, which provides further evidence that the swag helps promote our security awareness program and drive the desired behavior.

Swag may take on a life of its own. We typically print stickers for our security awareness campaigns because they are an inexpensive tool that we can get in high quantities. What has evolved in our organization is that numerous employees enjoy collecting our security awareness stickers. It is not uncommon to walk around the cubicles in our offices and see our stickers lined up on the employee's desk or overhead storage. This constant display increases our brand impressions, continually reminding viewers of our security program from this inexpensive product. Stickers offer an excellent opportunity to interact with your colleagues and build stronger relationships.

Shall We Play A Game?

Personally engaging with employees provides Security Operations with an opportunity to promote security awareness and build bonds with fellow employees that foster trust. We have found that greeting employees as they enter the building in the morning is an effective way to create a positive interaction. When running a campaign that includes stickers or Post-it notes as a promotional item, we seize the moment by stating the security message we are promoting and handing out the promotional item as employees pass by. This is a swift interaction, but it allows us to spread our security message to employees verbally and through our promotional materials.

We use this same type of interaction to play security games with employees. As employees approach our position in the building lobby, we ask if they would like to play a game for a prize. The majority of employees accept our offer to play one of the security games. The games are short, and the entire interaction typically lasts less than five minutes. The games are designed to reinforce our desired security behavior or to present an attack method that a malicious actor may use. An example of a game we play is asking the participant to identify which email address is potentially risky. We provide the participant with two sets of emails and inform them that each row contains one potentially malicious email account. We ask them to identify the risky account in each row and what makes it potentially dangerous.

Example

jsmith@bigenergy.com or jsmith@bigenerqy.com

jsmith@bigėnrėgy.com or jsmith@bigenergy.com

On the first row, the email on the right is potentially risky because the "g" in "energy" is replaced by a "q". In the second row, the email on the left uses Cyrillic or special characters, "ė", to impersonate the real domain.

These games are intended to create short, fun interactions between the employees and security operations while reinforcing techniques to equip the employees to be better defenders. The objective is not to stump the employee, but instead get them thinking about and understanding the various methods malicious actors may use to trick them into interacting with a harmful email. The prize for correct answers is typically a piece of our wearable swag, which helps increase participation, because our staff highly prizes the reward.

This has proven to be an extremely effective means of generating a positive engagement between security operations and our staff while providing quick security tips and training. An interesting side effect of these engagements is that we often have a line of participants who entered the building before our games started and have come back down to the lobby to participate. Through word of mouth, individuals in their department inform others that Security Operations is playing games in the lobby for swag. It's hard to imagine a scenario in which employees seek engagement with Security Operations, but that's what routinely happens when we host a game session. These sessions create a fun, positive interaction between Security Operations and employees while energizing the security culture.

Public Relations

Are marketing and public relations not the same function?

Marketing and public relations are often used interchange-ably. The confusion stems from the strong similarities between the two functions, which share complementary objectives and work towards a common goal: promoting a product/brand. The distinction between the two functions lies in the methods they use to promote the product/brand and the desired effect. Marketing strives to increase sales through promotion and advertising, while public relations focuses on establishing relationships with the public to maintain a positive brand image.

Public relations serves as a catalyst for developing a positive relationship between the organization's workforce and Security Operations. This employee networking effort fosters interest and knowledge in cybersecurity, thereby improving the overall culture. A key to developing this relationship is being genuine and honest with the employees to build a high level of trust.

Trust

Establishing trust between your employees and the Security Operations group is a primary benefit of fostering a strong relationship. Trust is a core element of any relationship, but building trust in a relationship is a long-term commitment. Developing your security awareness program from the perspective of "what is in it for the employee" will help foster this trust, as you will be helping them protect themselves while they help your security team protect the organization. Part of building this trust requires your Security Operations team to create events that allow them to interact with your organization's employees. If you are genuinely interested in helping them become more cyber-aware, your employees will believe in the program and trust your direction in protecting the organization.

Forging trust is a long-term investment that requires the Security Operations team to be consistent in their actions to nurture it. The way your Security Operations team responds to minor events, such as failing a simulated phishing test, will dictate how your employees perceive the relationship. If you are firm but understanding and take the time to work with employees to build their cyber skills, it will go a long way toward creating a very positive, trusting relationship. Conversely, suppose Security Operations belittle or ridicule employees when they make a mistake or ask a question. In that case, they may begin to distrust or avoid interacting with the Security Operations team, leading to a more adversarial relationship. This could lead to a strained or non-existent communication channel between the employees and Security Operations.

Trust is never more important than when an employee detects a possible security event. If you have fostered that strong sense of trust with employees, they will likely report the event immediately, minimizing any possible damage. Employees may hesitate or try to conceal a potential security event if they fear reprisal or do not trust the security operations team. During a cyber event, time is of the essence, and the sooner Security Operations becomes aware of a potential issue, the faster it can be addressed to avoid catastrophic consequences. This is where the time spent fostering these relationships can really help keep a minor cybersecurity incident from becoming a significant issue.

Trust must be built up over time, but can be lost in an instant, so handle it with great care.

Establish Employee Bond

Building and maintaining a strong relationship with our employees is critical to fostering a strong cybersecurity culture. Trust is one element of any relationship that cannot be taken for granted. Once you lose trust, it is almost impossible to regain. The relationship between Security Operations and the organization's employees relies on trust to establish effective, timely communication channels. When employees trust the security operations team, they are more likely to report potential incidents promptly.

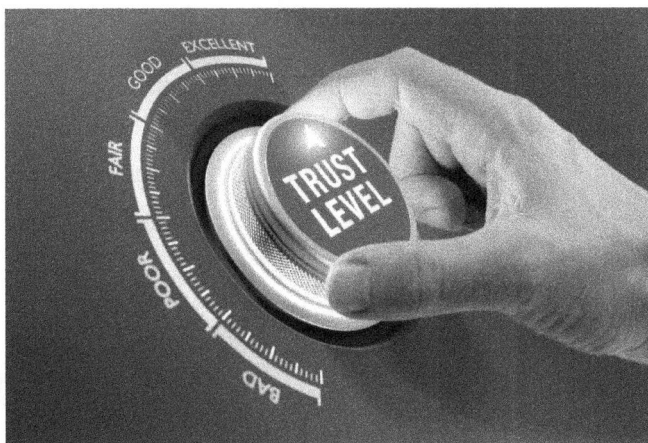

Figure 6.3: *Establishing trust*
Source: NicoElNino/Shutterstock.com

How does Security Operations build this trust with employees? Multiple facets go into building trust with employees, but I believe it begins by explaining what the role of Security Operations is within the organization. Security Operations is often viewed as the "police" within the organization, constantly monitoring the employees' activities. This can lead employees to have a negative view of Security Operations. If there is no positive relationship with security operations, employees may be hesitant to report potential security incidents for fear of negative consequences. Your security awareness program must create public relations campaigns that dispel this fear and build trust between your employees and the Security Operations team. We continually reinforce to our employees that Security Operations is not interested in any activity unless it poses a risk to the organization's digital assets. When we identify a potential risk, we will take appropriate measures to protect the organization's interests; otherwise, Security Operations remains in the background monitoring for potential threats.

The Department of Why?

Another way we establish trust with our employees is by working with them to actively identify mutually agreeable compromises that meet their business requirements and the organization's security requirements. Security operations is often referred to as the department of "NO." If you continually say "no," people will work around your security controls and go against the security culture you are trying to establish. This may lead to Shadow IT and the use of unauthorized applications, putting the organization's data at risk due to insufficient security controls. I prefer to refer to Security Operations as the department of "WHY." The reason for this is that I have found you get a much better response when you ask, "Why do you need access to the requested resource?" If the requester presents a solid business reason for the request, we can work on a compromise that provides access to the resource while maintaining an acceptable level of security.

This type of working relationship builds on that trust with employees that Security Operations is not trying to impede their business processes and will work with them to implement features that make them more productive in their jobs. This also helps prevent the use of "Shadow IT" as Security Operations can explain the risks, so employees understand the security reasons why something may not be approved for use.

Perspective

Approach your public relations from your employees' perspective. Find reasons or events to engage your employees and promote the adoption of good cybersecurity habits. If you show goodwill towards your employees, it helps build trust and a bond with your security awareness program, thereby driving a stronger cybersecurity culture. Create events that focus solely on improving or promoting the individual's security standing. If you help reinforce good security practices in employees' private lives, this behavior will become habitual and carry over to the workplace.

We held an event as part of an identity theft awareness campaign to encourage employees to freeze their credit. This event involved Security Operations handing out ice cream and information to help individuals understand the importance of not only freezing their own credit, but also their children's. A good portion of the employees who took our materials and listened to our advice had not thought about the ramifications of their children's credit being damaged. This was a great event that showed that Security Operations is invested in improving our employees' personal security. Remember, a key element of public relations is developing a positive impression of the brand through public interactions. There was no direct benefit to the organization in conducting this event, but it was an opportunity to demonstrate an interest and commitment to our employees' well-being. It also provided an excellent opportunity to interact with employees, build stronger relationships, and promote good cybersecurity practices.

Self-Promotion

Develop an excitement for your security awareness program through interactions with your fellow employees. When there is a sense of excitement, employees will be drawn to the program to learn more. The security awareness program lacks a voice, so Security Operations must become that voice, constantly promoting the virtues of participating in the program. Essentially, Security Operations must become the program's cheerleaders. This may be difficult for introverted individuals, but members of the Security Operations team must engage in public relations to foster relationships with the workforce. One of the best ways to ensure your employees remember your security awareness program is to tirelessly self-promote your product (improving the security culture).

Mattress Mack Master Promoter

If you are new to Houston, Texas, you will soon become familiar with the local icon, Jim "Mattress Mack" McIngvale. When he founded Gallery Furniture in 1981, he was anything but a household name in the Houston area. Turn the page forty years later, and you will be hard-pressed to find someone in Houston who does not know about Mattress Mack and Gallery Furniture.

He grew the Gallery Furniture brand through local television commercials featuring catch phrases such as "No, back, back backorder slip" or "Gallery Furniture saves you money!" Another element of these commercials was that Mattress Mack was frequently in the commercials wearing Gallery Furniture logoed items. This is a subtle but essential part of their market campaigns that constantly keeps their brand in the forefront while also making Mattress Mack one of the most recognized local figures in Houston. Anytime you saw Mattress Mack on television, he was continually promoting Gallery Furniture or other events he was participating in.

A key element of public relations is building strong relationships with the community. This is an area in which Mattress Mack has excelled and endeared himself to Houston and the surrounding communities through his philanthropic endeavors. Mattress Mack routinely supports numerous local charities and continually gives back to the community. His commitment to the community has included allowing displaced families to take refuge in his store following local disasters. These efforts have built significant goodwill between Houstonians, Mattress Mack, and Gallery Furniture. His commitment to public relations demonstrates the power of creating a strong public image and how it can facilitate positive brand awareness.

This tireless self-promotion of good security practices has several rewarding qualities. It is not uncommon for individuals to stop a member of the Security Operations team in the hallway to discuss a recent simulated phishing and jokingly inform us, "We almost got them." This is typically a lighthearted interaction, but the fact that the individual felt "safe" to stop a member of our Security Operations team to discuss this demonstrates a high level of trust built from our public relations campaigns. One of the most rewarding events occurs when an individual elaborates to a member of the Security Operations team about a phishing email or malicious text that the employee avoided on their personal devices. The employee is typically very excited, and you can sense their pride in themselves for avoiding the threat. These interactions reinforce the concept of promoting security practices in the employee's personal lives to establish good security habits that carry over to the workplace.

Summary

A strong cybersecurity culture cannot rely solely on Security Operations. It is crucial to build relationships with other members and departments in the organization to help promote your security initiatives. Utilize promotional items to generate interest and excitement within the workforce to gain participation in the cybersecurity program. Get out and meet with employees to build trust and promote the benefits of the security training provided.

Additional References

Influencers

Word-of-Mouth Marketing: Meaning and Uses in Business
https://www.investopedia.com/terms/w/word-of-mouth-marketing.asp

The Psychology of Celebrity Endorsements: Why We Trust Famous Faces
https://aaft.com/blog/advertising-pr-events/the-psychology-of-celebrity-endorsements-why-we-trust-famous-faces/

Influencer
https://sproutsocial.com/glossary/influencer/

The Power of Peer Influence in Marketing: Leveraging the Social Connection
https://medium.com/@alexanderizryadnov/the-power-of-peer-influence-in-marketing-leveraging-the-social-connection-df6ea6ec978c

The Marketing Psychology Behind Celebrity Endorsements
https://knowledge.wharton.upenn.edu/article/the-marketing-psychology-behind-celebrity-endorsements/

Swag

Promotional Products 101: A Guide to Using Them Effectively
https://www.ama.org/marketing-news/what-are-promotional-products/

Promotional Products Association International
https://www.ppai.org/

5 Ways to Extend the Lifespan of Your Swag
https://www.northpointprinting.com/post/5-ways-to-extend-the-longevity-of-your-swag

How the Beanie Baby Craze Came to a Crashing End
https://www.history.com/articles/how-the-beanie-baby-craze-came-to-a-crashing-end

Weirdest Collectible Fads of the 90s-2000s
https://go2tutors.com/weirdest-collectible-fads-of-the-90s-2000s/

Public Relations

Public Relations vs. Marketing for Business | University of Phoenix
https://www.phoenix.edu/articles/marketing/public-relations-vs-marketing-for-business.html

What is the Difference Between Marketing and PR?
https://www.meltwater.com/en/blog/what-is-the-difference-between-marketing-and-pr

Measuring Success

Measuring performance is a critical element in evaluating the effectiveness of security controls. This applies to the security awareness program, where the human element is the security controls being assessed. Gauging the success of your security awareness program and the adoption of your security culture is a daunting task. Security Operations uses numerous KPIs and metrics to assess the effectiveness of technological security controls, but assigning reliable values to evaluate the performance of human security controls remains elusive and challenging. Establishing metrics to measure the adoption of the security awareness program's desired behaviors provides insight into the effectiveness of the human security controls and the business justification for investing in the cybersecurity program. Earlier, I presented information to facilitate budget allocation for the security awareness program, so we must provide accurate metrics to support management's financial commitment to the cybersecurity program.

Risk Management

Security Operations uses metrics to determine whether deployed security controls effectively mitigate risk to an acceptable level. Unlike technical security controls, which may be replaced if deemed underperforming, the human controls are necessary for business operations and cannot simply be replaced when they are underperforming. We are working with human risk management, so, as with other security controls, when we identify a change in our risk profile, we must adjust our controls to mitigate it. Developing reliable metrics to assess the risk posed by the human element provides Security Operations with insights into trends that can help identify potential weaknesses in the security awareness program. Identifying gaps in security awareness training enables Security Operations to adjust the program to address those weaknesses, maintaining human risk at an acceptable level.

Establishing A Baseline

How can you measure the success of your security awareness program if you don't know where you are starting from? Once you have established the metrics you want to track and developed a reliable way to measure performance, take an initial sample to serve as your baseline. This baseline provides a starting point for monitoring your program's progress over time. Do not be discouraged by the initial sample. It is not uncommon for organizations conducting their first simulated phishing test to see a failure rate of over 30%. Once you have a solid baseline, you can begin formulating plans for your security awareness program to bring the measured value to an acceptable level.

Velocity Metrics

Velocity metrics are short-term metrics that measure week-over-week performance. These statistics tend to be volatile, often showing rapid or wider swings in the results when compared to neighboring samples. Tracking trends with velocity metrics is difficult due to the data variations. Velocity metrics are great for analyzing specific events in the security awareness training program, but may provide only limited indications of the program's overall performance.

Rolling Metrics

Rolling metrics are a foundational component for analyzing the data generated by the security awareness training program. The rolling metrics methodology continually observes a year's worth of data at every point of reference. Unlike velocity metrics, rolling metrics are designed to look at a larger dataset to evaluate trends in a program's performance. By using larger datasets, the program's statistics are presented in a smoother, more rolling manner, typically without wild fluctuations, allowing us to analyze trends and identify patterns of behavior over an extended period.

Let's review the rolling metrics methodology by examining a security awareness program that sends simulated phishing campaigns weekly. Each week serves as a marker for analysis from which statistics are calculated over the previous fifty-two (52) weeks, providing a richer view of the program's overall performance. The data points reference the previous fifty-two weeks (1 year), so the current data point analyzes the current year. In contrast, the oldest data point covers the prior year, providing two years of data. Each data point between the current and the last data point references the previous fifty-two weeks from that point, providing a very complete dataset.

Using **Figure 7.1** as an example, we can examine how the rolling metrics methodology works. In this example, we are currently evaluating data for the year 2025. Since we are evaluating a year's worth of data at every point, the start date for Week 1 of 2025 is 1/1/2024. As we move through the weeks of 2025, the start date advances, allowing us to analyze a year of data at each point. Upon reaching Week 52, we can evaluate all the data for 2025. Since Week 1's start date is 1/1/2024, we are essentially analyzing two years of data at this point. Reviewing the data in this manner enables a deep analysis of employee performance over an extended period. This expanded dataset provides a more accurate representation of the program's performance by accounting for employee performance improvements, technological shifts, and seasonal fluctuations in the organization's workforce.

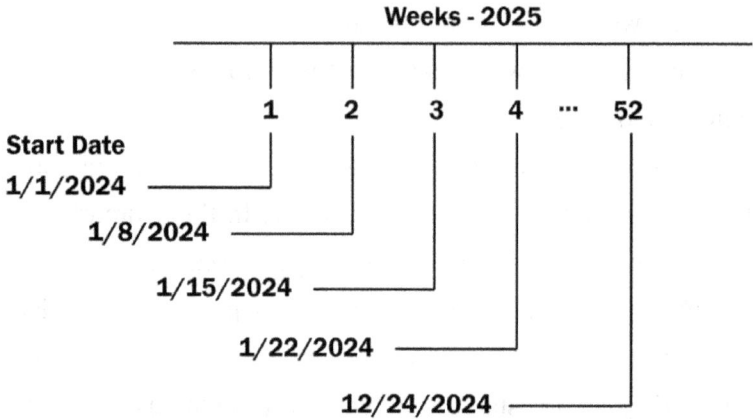

Figure 7.1: Rolling metrics methodology

Adjust the marker point in the rolling metrics to match the delivery schedule (weekly, bi-weekly, monthly, quarterly, etc.) of your simulated phishing campaigns.

What Are We Measuring?

The metrics should provide insight into the overall health of the organization's cybersecurity culture. This is derived by evaluating the performance and effectiveness of our security awareness training program to instill the desired security behaviors among employees, thereby reducing our human risk. The metrics used to measure the performance of our security awareness training are grouped into three categories: Click (failure) Rate, Reporting Rate, and Engagement Rate.

The Click and Reporting Rates are evaluated using velocity and rolling measurement methods. The Engagement Rate is a unique metric that uses rolling metrics to measure the number of users who engaged with potentially malicious content in our simulated phishing campaigns. Let's explore how these metrics can provide high-quality insights into the current human risk and the overall health of our security awareness program.

NOTE: As we begin our discussion on metrics, please note that the metrics referenced are based on a weekly cadence for the phishing simulation program. If your program is on a different cadence (e.g., bi-weekly, monthly, or quarterly), adjust the metrics to reflect your timetable.

Click Rate

The Click Rate (failure rate) indicates the number of failures in a simulated phishing campaign. Failure is defined as clicking a link, opening an attachment, or performing any other action that constitutes engagement with potentially malicious content in the email. Click Rate is often discounted by security professionals as a poor performance indicator that does not provide relevant data to assess the overall security culture. I disagree with this notion and believe the Click Rate provides very insightful performance data on the organization's cybersecurity culture when viewed through the right lens.

We previously discussed Tenure Groups, which group employees based on the length of time they have been in the security awareness training program. Incorporating the Tenure Groups into the delivery of simulated phishing campaigns greatly increases the value of the Click Rate statistic for assessing performance. The simulated campaigns for the Tenure Groups include several factors, including the maturity and size of the group, and the difficulty of the simulated emails, providing relevant differentiations and points of analysis between the groups. Analyzing the Click Rate by comparing the performance of the Tenure Groups allows for the identification of trends and adoption of the desired security behaviors.

A quick recap of Tenure Groups to assist in reviewing the sample charts:

Group 0 newest employees receiving the least difficult simulated emails

Group 1 more tenured group receiving more difficult simulated emails

Group 2 the most mature group receiving the most difficult simulated emails

Click Rates are evaluated using velocity and rolling metrics to provide an analysis of short-term and long-term performance. The metrics also provide reference points for assessing current performance against the defined campaign and strategic goals.

The following are sample Click Rate metrics we can use to understand how events affect our simulated email campaigns.

Click Rate – Velocity

The graph (**Figure 7.2**) shows Click Rate Velocity, a week-over-week performance comparison. The performance indicators show rapid fluctuations that provide some insight into performance trends, but those are not clearly seen. Click Rate Velocity is an excellent tool for examining why a particular group spiked in a given week. This may be traced back to a high failure rate on a specific simulated phishing email, or to a sudden influx of new members to a group (e.g., summer interns). This provides insight into short-term variations within the program, but does not present clear information about performance trends.

Energized Cybersecurity Culture

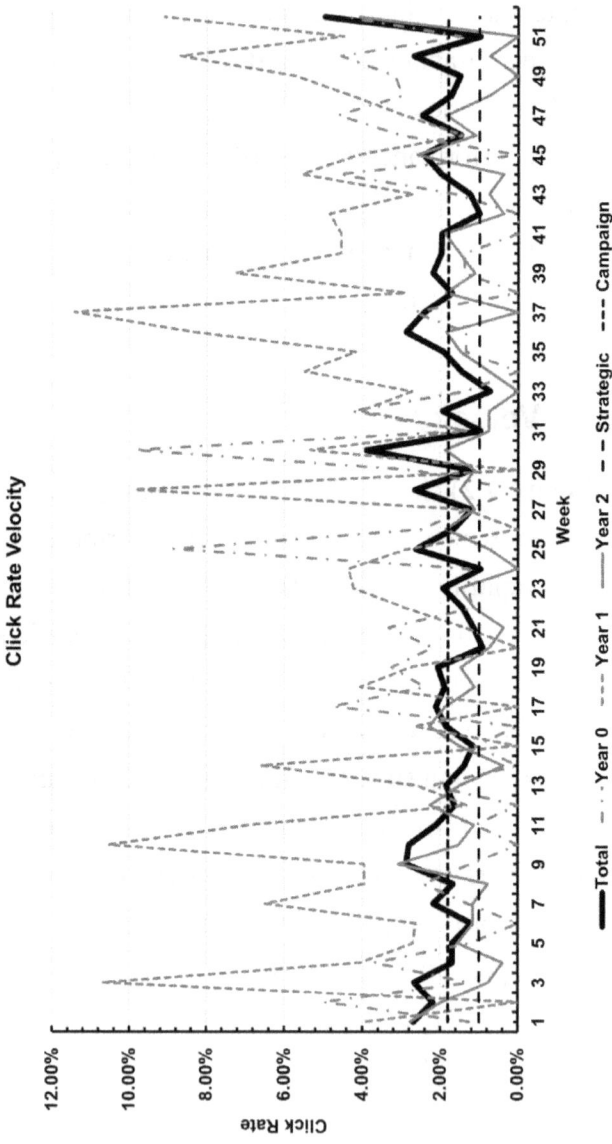

Figure 7.2: Click rate velocity

Click Rate – Rolling

Rolling Metrics provides a clear view of trends within the simulated phishing program and how changes within the security awareness training program are identified through this data. Analyzing this trend data allows us to adjust the security awareness training to improve our employees' results and the adoption of the desired security behaviors.

Reviewing the Click Rate Rolling Metrics for 2023 (**Figure 7.3**), there are several key takeaways from this data. The Year 2 group, which receives our most difficult emails, is the best-performing group, hovering around the defined strategic goal. The Year 0 group, our newest employees, has the highest average click rate, and the Year 1 group, which receives our mid-tier simulated phishing emails, is beginning to track more closely with Year 2. The phishing campaigns are designed to become more difficult the longer an employee is in the program, allowing them to build their skills to identify the characteristics of potentially malicious emails. The statistics show a progression of improvement from Year 0 to Year 1 to Year 2, with email difficulty increasing in each group. The data from the Click Rate Rolling Metric suggests that employees' ability to identify and avoid potentially malicious content improves as they progress through the security awareness training program. This supports the concept that our employees can become strong defenders and adopt the desired behavior by moving them through training that progressively challenges their abilities.

Examining the Click Rate Rolling Metrics for 2024 (**Figure 7.4**), it is apparent that something disruptive occurred in our simulated phishing campaigns, as the results for the Year 1 and Year 2 groups showed dramatic increases in click rate over a short period. These jumps in click rate were due to a technological change in the simulated phishing campaigns. The campaigns were adjusted in week 9 to begin sending AI-generated templates to Year 1 and Year 2 groups. The new AI-generated templates increased the difficulty by introducing simulated emails with content based on email threads the employees regularly participate in. Based on the dramatic increase in click rate, members in these groups appear to be struggling with the introduction of the more difficult AI-generated emails into the simulated phishing campaigns.

This provides evidence that we need to work with these groups to help them identify more sophisticated types of email threats. Based on the sustained increase in simulated phishing statistics, it was determined around week 35 that new training methods were needed to help our employees identify these threats. A new marketing campaign was initiated, and a new incremental campaign goal was set to bring our click rate back down. Over the next several weeks, the new training campaign was fully deployed, and a gradual downward trend was seen in Year 1 and Year 2 groups, providing statistical evidence that the new training was working. Security Operations should continue to monitor these groups to ensure performance continues to trend towards the defined goals. If new security technologies or training are provided, the campaign goal should be adjusted to reflect the expected results from these changes.

Energized Cybersecurity Culture

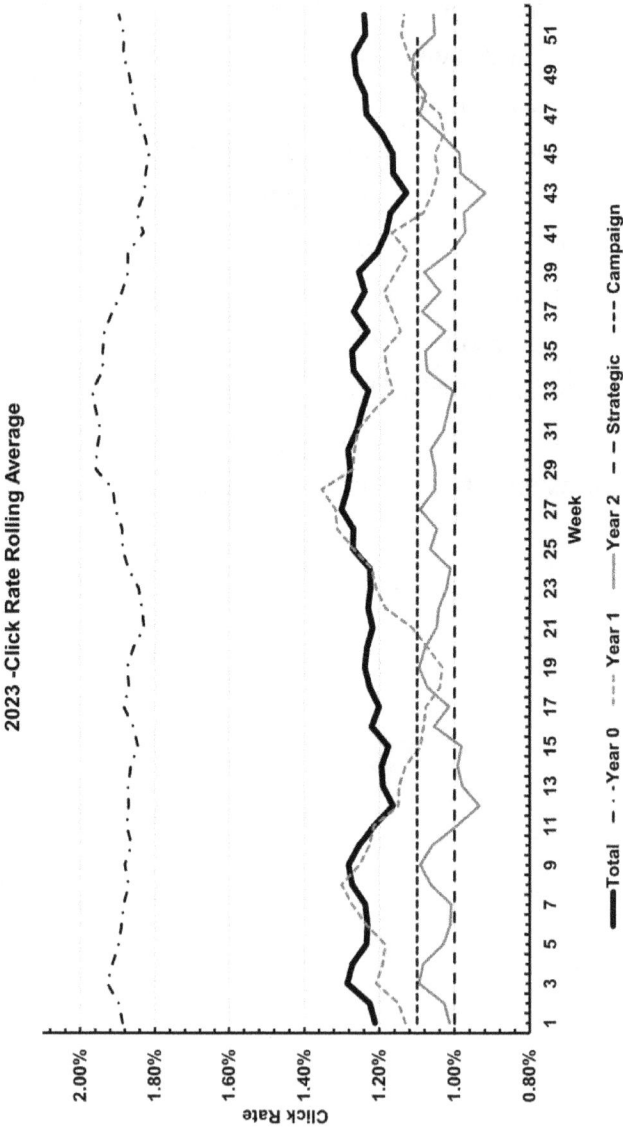

Figure 7.3: Click rate rolling metric - 2023

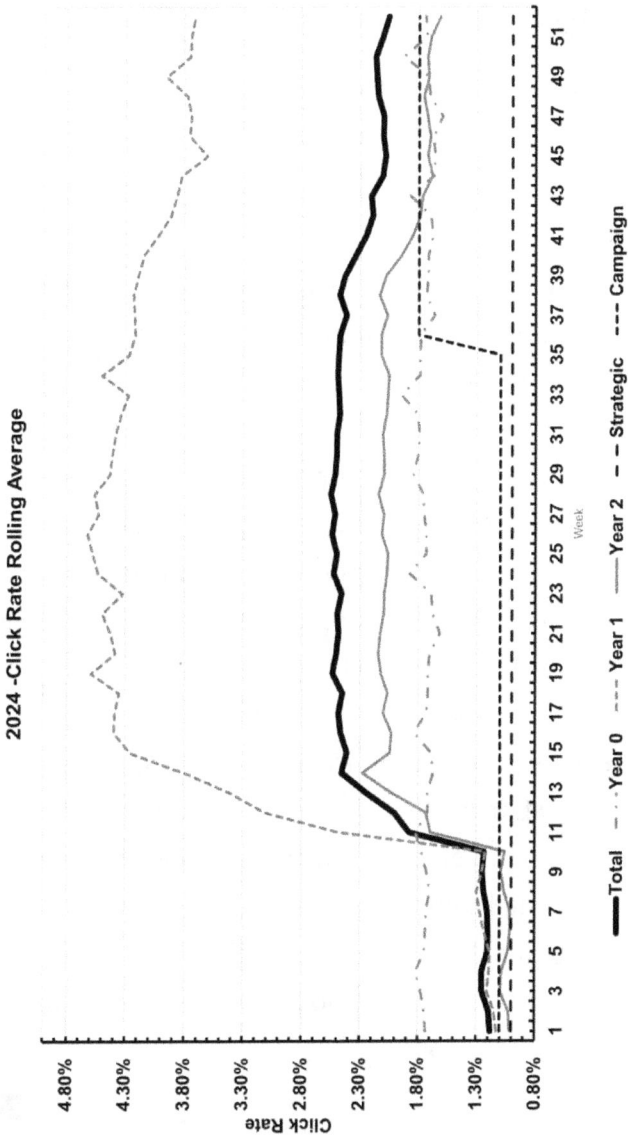

Figure 7.4: Click rate rolling metric - 2024

Reporting Rate

Many security professionals view a user reporting a potentially malicious email as the best behavior possible. Reporting Rate effectively measures the ability of your organization's employees to report emails that are part of your simulated phishing campaigns using the preferred reporting methods. Accurately measuring the Reporting Rate provides insight into the adoption and understanding of your desired security practices, assuming that employees take similar action when reporting suspect emails that are not part of the organization's simulated phishing campaigns.

The organization's security defenses involve multiple layers, and your employees are part of your security stack. Employees must be trained and encouraged to report potentially malicious emails to security operations for evaluation. Reported emails provide security operations with an opportunity to extract indicators that may be used to identify additional malicious emails or attacks against the organization. These indicators allow security operations to recognize employees who have received similar emails and remove them before any interaction with the email occurs.

Glorified Delete Button

It is essential to understand that Security Operations must walk a fine line when encouraging and discouraging users from reporting potential phishing emails.

The Reporting Rate, when accurately submitted, provides an effective metric to measure the ability of the organization's employees to identify and report suspicious emails. Many simulation platforms, such as KnowBe4, offer add-on products that embed a reporting button in the email interface, allowing users to easily report suspicious emails. The ease of reporting suspect emails may tempt some users to abandon reviewing the email's content and simply report marginally suspect emails. I term the overuse of the integrated reporting mechanism as the "glorified delete button." This is a condition in which an employee's mindset shifts to the belief that they are protecting the organization by reporting every piece of spam or junk mail in their inbox. This behavior has several adverse effects, including potentially wasting Security Operations resources for unnecessary analysis or skewing the reporting metrics.

Figure 7.5: *Glorified delete button*
Source: Me dia/Shutterstock.com

As part of the security stack, employees need training and reinforcement on the proper use of the phish reporting mechanism. This training should include standard techniques for identifying the characteristics of malicious emails and when to report emails that exhibit those characteristics. If users report every email without first analyzing its content, they have not entirely accepted their role in the organization's cyber defenses. Unnecessary reporting of emails negatively impacts the Reporting Rate metrics, as it does not accurately reflect the employee's ability to utilize their skills and training to identify a potentially malicious email. Identifying and providing additional training for users who misuse the phish reporting mechanism helps strengthen overall network defense, reduces unnecessary use of Security Operations resources, reinforces the organization's cybersecurity culture, and improves the effectiveness of the Reporting Rate metrics.

Understanding Reporting Rate

Organizations typically deploy an array of email defenses to analyze and block email before it reaches a user's inbox. This does not prevent all suspect emails from reaching their intended destinations, as malicious actors continually evolve their tactics to evade email defenses. We need our employees to actively defend their inboxes by reporting any emails they deem potentially malicious.

Let's evaluate some sample Reporting Rate metrics and determine what we can learn from them.

The Reporting Rate Velocity Metrics (**Figure 7.6**) displays a typical chart for the organization's yearly reporting. The charting of this information continues the use of our tenure-based groups (Year 0, Year 1, and Year 2), which provides valuable insight regarding the organization's overall adoption of the desired security behavior. Based on the information in the chart, the Year 1 and Year 2 groups follow a similar pattern, indicating that they have adopted a similar behavior. At the beginning and end of the chart, these groups are performing close to the desired campaign and strategic goals, but their performance dips around week 8. The reporting pattern may be due to the introduction of a new technology that required an update or to a change in the location of the reporting mechanism.

The Reporting Rate Rolling Metrics (**Figure 7.7**) effectively demonstrates that as users progress through the security program, they begin to exhibit the desired reporting behavior. The most mature group, Year 2, consistently scores near the defined goal levels. Year 1 closely follows Year 2, the top tier, providing evidence that as users advance through our security awareness training program, they are adopting the desired reporting behavior. Our newest employees, Year 0, are well below the goal levels but remain consistent in their performance. The information provided by this graph supports the idea that as users progress through the organization's security program, they begin to adopt the desired security behavior and the underlying security culture.

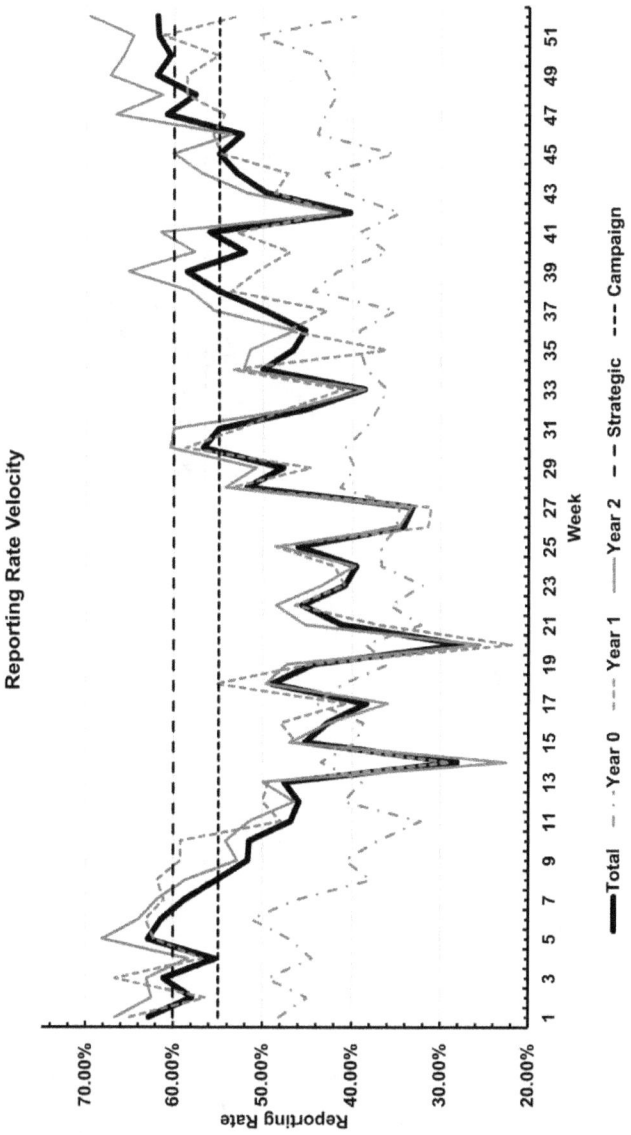

Figure 7.6: Reporting rate velocity

Energized Cybersecurity Culture

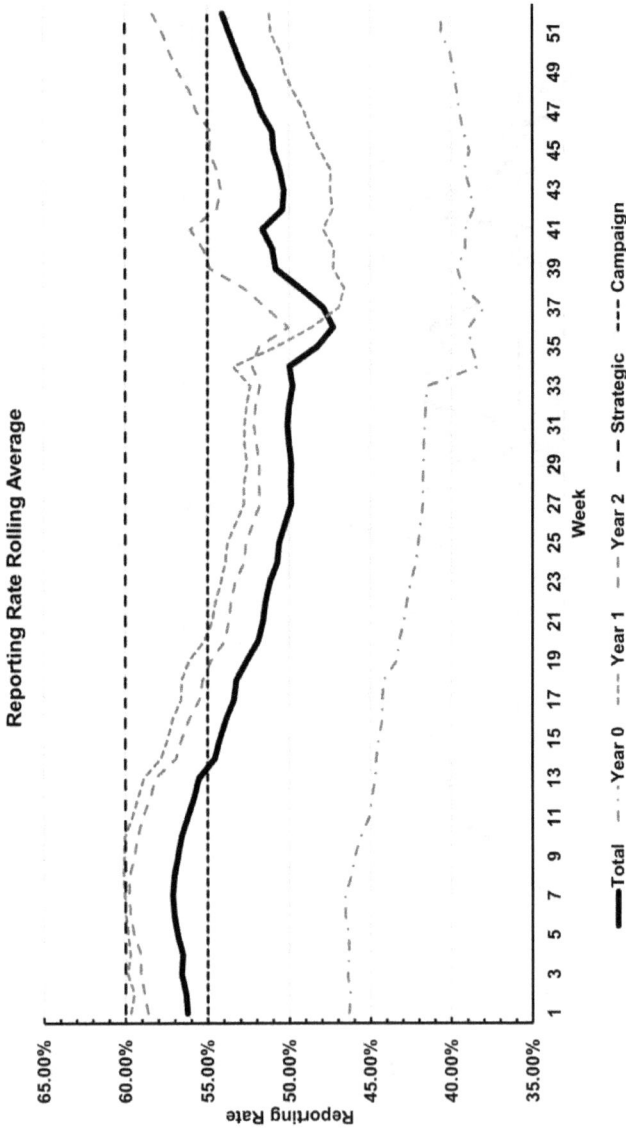

Figure 7.7: Reporting rate rolling metric

Engagement Rate

Reporting an email may be viewed as the absolute best course of action, but what happens if the employee deletes a suspect email rather than reporting it? This is where tracking the Engagement Rate becomes necessary to paint a clear view of the organization's human security defenses. Engagement Rate is a key metric to evaluate employee adoption of the organization's desired action and performance in protecting the network from potential email threats. The Engagement Rate analyzes user interaction with the content in simulated phishing campaigns. What makes the Engagement Rate a valuable metric is its ability to measure how many users engage with potentially malicious content, such as links, attachments, or other calls to action, within the simulated phishing emails. We previously looked at the Reporting Rate, and I believe that employees reporting a suspected phishing email is the best response. This action prevented the user from engaging with the email's content and provided security operations with the opportunity to identify valuable indicators that may expose other possible attacks.

What about the users who delete the email? These users didn't engage with the email's content, so doesn't this action also demonstrate a desired response?

Who Forgot to Turn On Logging?

Employees are often referred to as "Human Firewalls," indicating they are a layer in the security stack. Traditional firewalls protect the edge of the network from connections and threats originating on the Internet. Just like their technological counterparts, the human firewalls are also responsible for protecting the edge of the network from online threats.

Let's examine a scenario by looking at a traditional firewall and how it is commonly configured. Firewalls typically block or allow network traffic based on policies, and we can often determine whether these policies are working by reviewing the firewall logs. While designing a new firewall policy to block some undesirable content, we forgot to enable logging. When we begin examining the firewall logs, we do not see any indicators that this traffic is being blocked by the firewall, even though the policy is configured to block it. Since we cannot see traffic being blocked in the firewall logs, does that mean the policy is not performing the desired action? The policy is working as expected, but we do not see the results to give the policy credit for blocking unwanted traffic.

What if we applied this same logic to our employees, who delete potential phishing emails without reporting them?

They are blocking the unwanted threat by deleting the email, but since they are not reporting it, they do not receive the credit for taking an action that protects the organization. This is essentially the same scenario as the firewall policy that blocks undesirable traffic but is not set up to log the action.

Deleting a suspect email is the second-best action an employee can take because it avoids engaging with any potentially malicious content it may contain. By deleting the email, Security Operations potentially missed an opportunity to obtain indicators, but the employee did perform the primary desired action of preventing an email threat from entering the organization's network. Our security awareness program aims to train employees to be active participants in the organization's cybersecurity defenses. Deleting suspect email demonstrates the employee's behavior is consistent with defending the network, but, as with the misconfigured firewall policy, neither will receive acknowledgment for their actions.

The Engagement Rate metric attempts to fill this "no logging" gap when users neither report emails nor interact with the content. Deleting an email is an acceptable action, but how can we measure this? The Engagement Rate creates a metric based on a user's interaction with our simulated phishing campaigns to determine if they engage with any of the content.

Let's Do Some Math

The Engagement Rate (see **Figure 7.8**) enables analysis of how many users interact with malicious content in our simulated phishing campaigns. This provides insight into the adoption of desired security behaviors and employees' ability to avoid threats. The Engagement Rate is a perpetual metric that follows the Rolling Metric methodology.

Engagement Rate is calculated by counting the number of unique users who engaged with the content in simulated phishing emails during a given timeframe. It is important to note that there is special consideration for how we "count" users who interact with a simulated phishing email. Count represents the number of unique users who engaged with at least one simulated phishing email. A user who interacts with more than one simulated phishing email within a specified timeframe is counted only once. We do not count each failure, as we are only trying to identify individuals who have engaged with the content. Another way to look at it would be as a Boolean. Did the user engage with the content, yes (1) or no (0)?

Count - number of employees tht interacted with one or more simulated phishing emails during the timeframe

Average Number of Employees - average number of employees that received simulated phishing emails during the timeframe

$$Engagement\ Rate = \frac{Count}{Average\ Number\ of\ Employees}$$

Engagement Rate is calculated as the Count divided by the Average Number of Employees in the specified timeframe. This effectively gives us a percentage of employees who engaged with content in our simulated phishing campaigns for the specified timeframe.

Energized Cybersecurity Culture

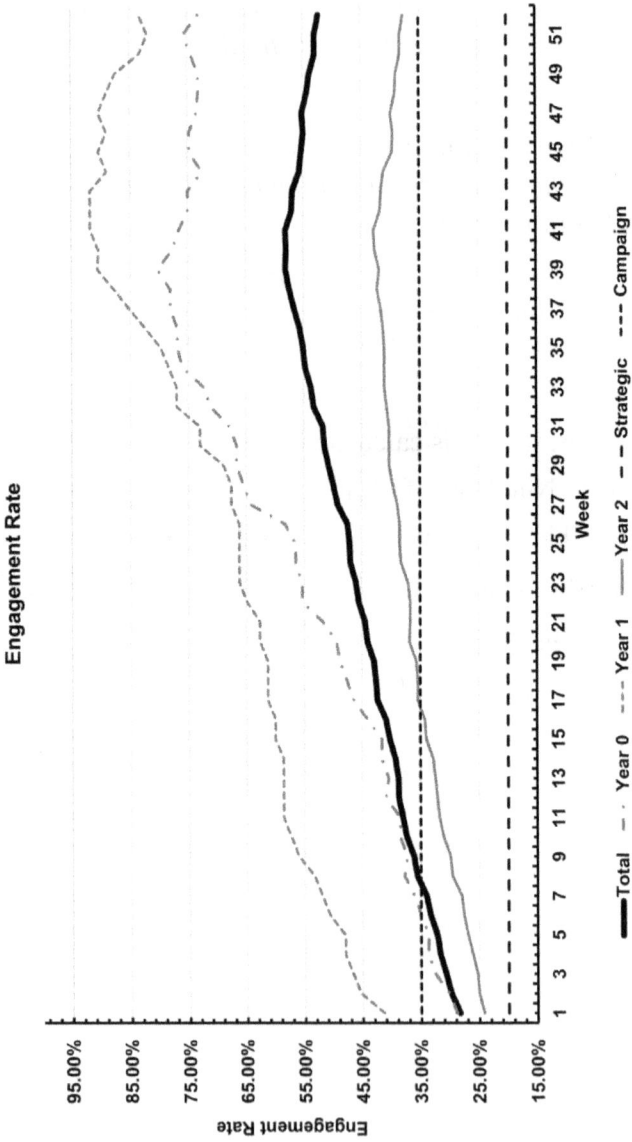

Figure 7.8: Engagement rate

Reverse Market Share

We are examining how to enhance our security awareness program by incorporating marketing techniques. A common goal of many marketing strategies is to grow our product's market share by increasing the number of people using our product. As cybersecurity practitioners, we want to increase our employees' participation in our cybersecurity culture to expand our customer base and overall market share. Market share is identified by the percentage of a product's sales in comparison to the total sales for the market.

Analyzing your cybersecurity culture from a market share perspective, you can use the total number of employees in your organization as the total market size, and the market share of your program is defined by those employees who routinely refrain from engaging with potentially malicious content. Individuals in our organization who interact with one or more of our simulated phishing emails demonstrate they are not performing our desired security behavior. We identified this group of users when we created the Engagement Rate (Figure 7.8: Engagement rate), which identifies users who engage with our simulated phishing campaigns. Given that this group of users does not exhibit our desired security behavior, we can infer that the Engagement

Rate represents individuals who are not part of our market share. Another way to look at his measurement is that the inverse of the Engagement Rate represents our current market share, thus the "Reverse Market Share".

Engagement Rate = 48%

Current Market Share = 1 - .48 = .52 or 52%

The goal of your simulated phishing program is to get the highest number of users to successfully identify and report the email, or at least not engage with the email's content. Users who repeatedly perform this behavior have adopted your security culture and represent the program's market share. Users who do not routinely perform the desired behavior have not adopted all the techniques presented in your program, representing potential market share. To bolster your market share and adoption of your cybersecurity culture, you need to identify users in this potential market share and develop new methods to communicate and deliver your message about your cybersecurity program. As members of this potential market share begin taking the right actions regularly, your cybersecurity culture will become stronger, thereby growing your market share. Dominating the market (culture) is a common goal of marketing campaigns and strategies, and we achieve this domination by increasing the number of users who adopt and perform our desired security best practices.

Scorecards

Scorecards are an effective tool for quickly providing participants in your program with feedback on their performance. The scoring system I have developed and incorporated into my security awareness program is based on our simulated phishing campaigns and training. The scoring is calculated on a one-hundred (100) point scale that includes the number of failures, weeks since last failure, number of reported emails, and the number of weeks past due on assigned training. An additional weighting factor is multiplied by the score based on the user's position. There is no standard or correct formula for creating your scorecards. When developing your scoring system, identify the measurable aspects of your program and assign a weight to each measurement. When selecting measurements for your calculation, consider metrics that reflect the behavior you are trying to encourage.

A key element of the scorecard is to represent their performance in a simple, widely recognized manner that requires no additional explanation. My workforce is based solely in the United States, so I choose to use the letter grade scale (A – F) commonly used in schools. This allows me to display grades in a single-character format that my workforce can easily understand. If you are creating a scorecard for a multinational organization, you will need to identify a simple, universally recognized score. Simplicity is an essential component of the scorecard, as you want your users to quickly and easily identify their performance.

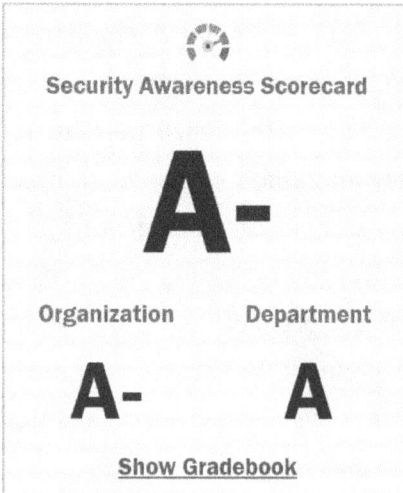

Figure 7.9: Security Awareness Scorecard

The scorecard is delivered to employees via the organization's intranet (Figure 7.9: Security Awareness Scorecard), and displays the scores for the organization and the employee's department. This provides employees with immediate feedback on how their performance compares with that of their peers in their department and across the organization. This creates an element of gamification driven by competition within the workforce, as employees do not want to be seen as underperforming compared to their peers. The desire of employees to improve their grade drives them to adopt the organization's preferred security behavior. Including department and organizational grades increases department leaders' participation in driving employee performance improvement, as they do not want their departments viewed as underperforming.

Gradebook

The Scorecard includes a Gradebook view that provides a breakdown of the scoring. This allows the employee to quickly identify which behaviors they need to improve to meet the organization's expectations. If the employee has no subordinates, only their grades will be displayed. Supervisors receive a complete Gradebook to quickly review their subordinates' scores and overall grades. This view also provides a breakdown of any departments or groups under the supervisor. This continues up the organization chart to the CEO, who can view the gradebook for the entire organization.

The Gradebook provides another opportunity to gamify the results of the security awareness training program, thereby improving employee performance by ranking scores across all departments. This provides an element of competition between departments, much like that among individuals, where supervisors do not want their department identified as underperforming compared to others. This helps create a top-down push, as supervisors, in an effort to maintain or improve their department's grade, will encourage employees to enhance their individual grades by performing the desired security behavior.

Security Awareness Gradebook

Organization Department

A-   A

Employees

| | | Scores | | | | | Emails | |
Employee	Grade	Total	Phishing	Weeks	Reporting	Training	Sent	Failures	Reported
Bill Smith	A-	91.0	54.0	4.0	3.0	30.0	49	1	37
Amy Johnson	A+	99.0	60.0	5.0	4.0	30.0	50	0	44
Nick Sims	A+	98.0	60.0	5.0	3.0	30.0	50	0	36

Scores Legend

Phishing - (60 points) lose points for each failure
Weeks - (5 points) points since last failure
Reporting - (5 points) based on number of emails reported (50% minimum)
Training - (30 points) lose points for each week assigned training is past due

Figure 7.10: *Security Awareness Gradebook*

Summary

Metrics play a vital role in understanding how well your employees are adopting the desired security behaviors. There are several standard metrics for measuring the security awareness program's performance, but look for other ways to use the data to create meaningful metrics for your environment. Metrics should not be limited to Security Operations but expanded to provide feedback to organizational leaders to help foster the cybersecurity culture.

8

Final Thoughts

The premise of this book was to present some ideas and methods I have found to be effective during my journey in creating a strong security awareness program and associated cybersecurity culture. The intention was not to create a how-to guide, but rather to provide content and ideas that can be adapted and customized to work in your program. There is no magic formula for creating a successful security awareness program or improving the organization's cybersecurity culture. Each organization is unique; it is up to the security practitioners to identify and foster a strong cybersecurity culture. Hopefully, the content presented will spark some ideas that facilitate improvements in your security awareness program, ultimately leading to a strong cybersecurity culture within your organization.

Human risk management is precisely that: managing the risk associated with the human element. The human element is one of the most attacked controls in our security stack, yet it often receives the least attention. Approach the management of human risk the same as you would for any other technical control in your security stack. Understand the cost and return on investment to develop a budget to facilitate mitigating this risk to an acceptable level. The challenge in mitigating the risk associated with human control lies in the fact that this control consists of numerous individual controls, each functioning slightly differently from the others. To meet this challenge, you must develop a patch management program for the human element that is adaptable and can facilitate the coordinated effort of these individual controls to protect your organization.

Incorporate marketing techniques into your security awareness program to manage and reduce your human risk. Develop a brand for your program and start building brand awareness among your workforce. Approach the advancement of your security awareness program with the mindset that you are trying to increase your market share. Improving brand awareness and growing market share are common goals of marketing strategies. Identify your program's current status and develop a marketing strategy that outlines the goals for your program over the next several years. Identify your target audience, create a budget, and develop campaigns to increase awareness and engagement with your security awareness program to achieve the defined goals. Measure your progress by analyzing key performance indicators.

The strategic goal for your security awareness program should focus on educating your workforce and instilling the importance of the employee's role in the overall defense of the organization. Identify the makeup of your target audience and how you plan to present the value of actively participating in the cybersecurity program to this group. Establish a baseline based on your organization's current state of cybersecurity culture and develop a marketing strategy that outlines the long-term plans and goals for your security awareness program. The goals defined in your marketing strategy will likely include creating brand awareness for the security program and advancing the program's reach through increased market share. Increasing market share will be achieved by modifying employee behavior to align with the desired behavior of your security awareness program, which directly enhances the organization's cybersecurity culture. Develop smaller marketing campaigns that connect with your employee base and foster progress towards the goals of the marketing strategy.

Marketing your security awareness program can be challenging because we must find ways to engage with individuals whose foremost thought is not cybersecurity. Develop the security awareness program from the employee's perspective, highlighting the benefits it offers and marketing the desirable lifestyle that comes with active participation. Develop a structure that allows you to classify groups of your workforce to deliver appropriate training materials and serve as a reference for establishing your metrics. Create simulated phishing campaigns that are relevant to the threats faced today and increasingly challenge the employees' skills. Video training campaigns must be engaging to attract and retain employees' interest. Your audience is diverse, so experiment with different types of content to appeal to all the members of your workforce.

The development and implementation of your security awareness program are among the few security controls where you can be creative and have fun. Venture outside the box and find innovative ways to engage with your workforce, helping to promote the reasons why individuals should adopt the desired security behavior. Establish a network of influencers to help spread the word and advance the positive effects of participating in the security awareness program. Utilize promotional items to reward desired behavior and draw attention to participation in the security awareness program. Establish a connection between the security program and the organization's employees that is based on trust. When employees trust the program's benefits, they are more likely to want to participate and adopt the desired secure behavior. Trust is fragile and requires continual nurturing.

Improvements in employee behavior related to the security awareness program often translate into improvements in the overall cybersecurity culture. Analyze the measurable variables in the security awareness program and identify those variables that are important to your organization. Develop ways to measure the identified variables and establish a baseline to measure progress. Measure changes in performance from the baseline to demonstrate improvement in your employee's behavior in support of the desired cybersecurity culture. Attempt to gamify aspects of your metrics to create friendly competition within your organization, thereby improving the overall security behavior of the organization. Identifying weaknesses in the security awareness program is crucial for making the necessary corrections to establish a robust cybersecurity culture.

Energized Cybersecurity Culture

This book chronicles my journey in developing security awareness training programs, which have helped improve the organization's cybersecurity culture. I hope some of the knowledge I gained during my exploration can help you facilitate a strong cybersecurity culture within your organization. My adventure continues, so as you embark on your journey, understand that there is no end. Your workforce is constantly in flux as employees leave the organization and new employees join. Additionally, the threat landscape is continually evolving, so you must constantly develop new ways to update your employees' firmware. Regardless of your workforce's maturity, your security awareness program needs to continually introduce new ideas and materials to properly educate your workforce and keep them highly engaged in your cybersecurity culture. Look for ideas from a variety of sources to help keep your security program relevant, informative, and engaging for your workforce. If you neglect the security awareness program and cyber-security culture, you risk losing a substantial component of your security stack.

The biggest compliment I get from individuals who have participated in my security awareness program is that I make learning about cybersecurity fun. When you make something fun, people enjoy participating in the endeavor.

Approach your security awareness program from a marketing perspective to engage your employees and generate enthusiasm for the security awareness program. Be creative, interact with your employees, and find fun, unique ways to generate excitement for the program, fostering increased participation.

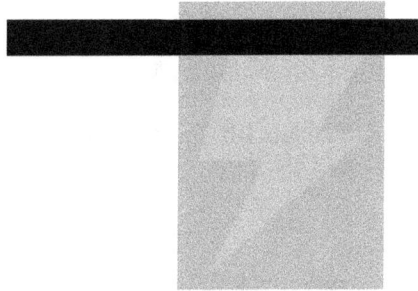

Appendix A:
Security Awareness Dates

Data Privacy Week

Dates vary, but always around January 28th

National Identity Theft Awareness Week (FTC)

Date varies, last Monday in January

Safer Internet Day

Date varies, usually the first week in February

World Backup Day

March 31st

Identity Management Day

Second Tuesday in April

World Password Day

First Thursday in May

National Social Engineering Day (Kevin Mitnick's Birthday)

August 6th

Cybersecurity Awareness Month

October

International Fraud Awareness Week

Usually, the second week in November

Computer Security Day

November 30th

National Cookie Day

December 4th

Look for other events/days that can be related to cybersecurity topics.

Appendix B:
Additional Resources

National Cybersecurity Alliance

https://staysafeonline.org/

Federal Trade Commission

https://bulkorder.ftc.gov

SANS

https://www.sans.org/

KnowBe4

https://www.knowbe4.com

Hoxhunt

https://hoxhunt.com/

ProofPoint

https://proofpoint.com/

Amazon

https://learnsecurity.amazon.com/en/index.html

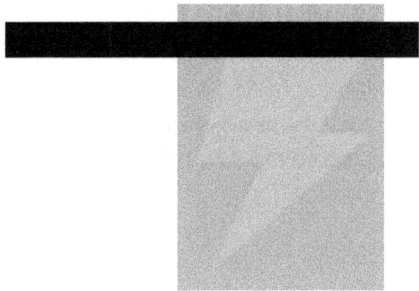

Index